ROUTLEDGE LIBRARY
20TH CENTURY SCI

Volume 8

RECENT DEVELOPMENTS IN
ATOMIC THEORY

RECENT DEVELOPMENTS IN ATOMIC THEORY

LEO GRAETZ

Translated by
GUY BARR

Routledge
Taylor & Francis Group

LONDON AND NEW YORK

First published 1923 by Routledge

2 Park Square, Milton Park, Abingdon, Oxon OX14 4RN
711 Third Avenue, New York, NY 10017, USA

First issued in paperback 2016

Routledge is an imprint of the Taylor & Francis Group, an informa business

Notices
Practitioners and researchers must always rely on their own experience and
knowledge in evaluating and using any information, methods, compounds, or
experiments described herein. In using such information or methods they should
be mindful of their own safety and the safety of others, including parties for whom
they have a professional responsibility.

Product or corporate names may be trademarks or registered trademarks, and are
used only for identification and explanation without intent to infringe.

British Library Cataloguing in Publication Data
A catalogue record for this book is available from the British Library

ISBN-13: 978-1-138-01355-1 (hbk)
ISBN-13: 978-1-138-98471-4 (pbk)
ISBN-13: 978-0-415-73519-3 (Set)
eISBN-13: 978-1-315-77941-6 (Set)
eISBN-13: 978-1-315-77932-4 (Volume 8)

Disclaimer
The publisher has made every effort to trace copyright holders and would
welcome correspondence from those they have been unable to trace.

RECENT DEVELOPMENTS IN ATOMIC THEORY

BY

LEO GRAETZ

TRANSLATED BY

GUY BARR, B.A., D.Sc.

WITH THIRTY NINE ILLUSTRATIONS

METHUEN & CO. LTD.
36 ESSEX STREET W.C.
LONDON

First Published in 1923

PRINTED IN GREAT BRITAIN

PREFACE TO THE GERMAN EDITION

THE subject-matter of these lectures, which were delivered in different places, some of them in the territory occupied during the war, is, as I have been able to satisfy myself, of general interest, not only to students of physics and chemistry but to the majority of people of a scientific turn of mind. It seemed therefore to be worth while to make these lectures available to a wider circle, especially as I have striven in them to make clear the steps by which the recent revolutionary views about atoms have been obtained, and to illustrate the advances which have been made by their means in the explanation of many phenomena. The fact that these new conceptions are still incomplete, and still present various difficulties and obscurities which can only gradually be removed, does not in the least affect their value, but only shows that science has here penetrated into a new and unknown but fruitful territory.

MUNICH, *August*, 1918.

The fourth edition has been enlarged, particularly by a discussion of the recent brilliant investigations of Aston on isotopes and of Rutherford on the structure of nuclei.

MUNICH, *February*, 1922.

TRANSLATOR'S PREFACE

VERY few deviations have been made from the fourth German edition. I have not hesitated, however, in using nomenclature to which English readers are accustomed, in preference to a simple transliteration of the German equivalent, in cases of divergence. The list of isotopes at the end of the third chapter has been extended in accordance with the later work of Aston and Dempster. Some sections of the sixth chapter, dealing with the nuclei of range 7-9 cms. which Rutherford found among the products of disintegration of nitrogen and oxygen and designated provisionally as X_3, have been omitted[1] in view of his recent statement that these particles appear to have their origin, not in the atoms of the gas, but in the source of a-rays used for bombardment.

GUY BARR

December, 1922

With the concurrence of the author—L. G.

CONTENTS

CONTENTS

RECENT DEVELOPMENTS IN ATOMIC THEORY

LECTURE I

MOLECULES AND ATOMS IN CHEMISTRY AND THE KINETIC THEORY OF GASES

THE early philosophers realized that, in spite of appearances, it was a fallacy to suppose that a piece of iron or glass completely fills the space which it occupies, leaving no gaps. The circumstance that all bodies contract on cooling so as to occupy a smaller space and that their volume may be reduced by pressure, shows that they could not have filled up the space before. But the gaps existing in bodies which are apparently continuous must be very small, for we have no indication of them even when we apply the most extreme magnification afforded by modern microscopes. Matter must therefore be composed of structural units, which are so small as not to be recognizable under the microscope, and these must be so assembled as to leave between them gaps of which the extent may be increased or reduced, e.g. by variations of temperature or of pressure.

This general idea has received the most complete confirmation, and has been rendered more definite

I

by the facts accumulated by chemical research. The enormous number of chemical compounds which either occur naturally or have been prepared in the chemist's alembic, always show certain definite relationships of weight between the ingredients of which they are composed. The simple and sufficient explanation of this fact, which has been the foundation of the whole of chemistry, is that every simple *elementary* chemical substance consists of very small particles or *atoms*, each possessing a certain mass : all the atoms of one substance are similar and have the same mass, but the atoms of different substances differ both in their mass and in their properties. There are about ninety chemical elements ; consequently there are about ninety different atoms. By the combination of different atoms there are formed the *molecules* or smallest units of all the chemical compounds. The actual mass in grammes of any of the ninety atoms can now, in fact, be stated : we shall cite the figures later. The *atomic weight* is to be understood, however, not as the actual mass of an atom in grammes, but as the ratio of the mass of that atom to the mass of an atom of the lightest substance, hydrogen. Atomic weights are thus merely relative numbers : they are mostly determined by chemical means with the aid of the balance.

The chemical symbols for the different elements, H for hydrogen, He for helium, O for oxygen, N for nitrogen, Li for lithium, etc., connote not merely the names of the elements but also their atomic weights. Thus N stands for the name nitrogen, and at the same time for the number 14, which is its atomic weight. A *gramme atom* of a substance is that number of grammes which is given by the atomic weight, e.g. a gramme atom of nitrogen is 14 grms.

Bodies which are not elementary, but compound, are formed, as we have seen, by the close union of one or more atoms of different elements : there is thus produced a small portion which we call a *molecule* of the substance. Molecules are the units of structure of compounds. They may be separated into their atoms by chemical agencies. The *molecular weight* is always equal to the sum of the atomic weights of all the atoms contained in the molecule.

It would thus appear that there is a difference between elementary and compound bodies, e.g. between the element, chlorine, and the compound, hydrochloric acid, in that the former are made up of atoms, the latter of molecules. It may be shown, however, that the individual units of construction, even of elements, are generally not separate atoms, but that two or more atoms are again closely united to form a molecule, consisting in this case, however, not of atoms of different kinds, but of similar atoms. This statement is a consequence, at any rate, for the gaseous state of matter, of *Avogadro's Hypothesis*, an hypothesis which is not capable of direct and rigid proof, but from which the deductions are always found to be true. Gases and vapours, whether they be gaseous elements such as hydrogen, chlorine, sulphur vapour, etc., or gaseous compounds such as hydrochloric acid, ethane, carbon dioxide, whatever the complexity of their molecules, all follow the same laws and show the same dependence of volume upon pressure and temperature : they must, therefore, have something in common in spite of the differences in their atomic or molecular weights. This common property was recognized in the rule stated by Avogadro that "*equal volumes of gaseous bodies at the same temperature and pressure contain the same number of molecules.*" Now, experiment shows that a litre of hydrogen and a litre of chlorine combine to give 2 litres of hydrochloric acid gas, containing 1 atom of hydrogen and 1 atom of chlorine to the molecule. If the molecules of the 1 litre of hydrogen and the 1 litre of chlorine contained only 1 atom each, it would only be possible for them to produce 1 litre of hydrochloric acid. Therefore, a molecule of hydrogen must consist of 2 atoms of hydrogen and a molecule of chlorine of 2 atoms of chlorine. The two similar atoms of hydrogen must, therefore, have been closely combined to form a molecule, and the same applies to the two similar chlorine atoms. The why and the wherefore of this is, however, an open question.

We have been able to prove that the molecules of most elementary substances consist not of 1, but of 2 atoms. But in the case of mercury vapour it has for some time been thought, for chemical reasons, that its molecule contained just 1 atom of mercury, i.e. that it was a monatomic molecule. Now, for a monatomic gas, the Kinetic Theory of Gases makes a certain prediction. Two different specific heats are distinguished in the case of gases, the specific heat at constant pressure (c_p), and that at constant volume (c_v). The Kinetic Theory of Gases shows that for a monatomic gas the former is 5/3 of the latter, that is, the ratio of the two specific heats must be 5/3. This prophecy has been strikingly fulfilled by the well-known measurement by Kundt and Warburg of the ratio of the two specific heats for mercury vapour. The monatomic nature of mercury vapour was thus demonstrated physically. For a long time mercury was the only substance known of which a molecule contained only 1 atom. But when Ramsay discovered the new gases in the atmosphere, the so-called rare gases, argon, neon, krypton, xenon, and helium, it was found that they all gave again the value 5/3 for the ratio of the specific heats. They also must, therefore, be monatomic, and for some reason or other closer combination of 2 atoms does not occur and appears to be impossible. Further, these elements are such that they do not combine even with the atoms of other elements ; they are inert elements, their atoms always persist in perfect isolation.

With the exception of the monatomic gases, then, all bodies are composed of molecules. But molecules are made up of atoms. The atoms of the elements are, therefore, the ultimate smallest units of all molecules and so of all bodies. As we have seen, there are about ninety different elements. There are, therefore, about ninety different units, the combination of which produces all material bodies, whatever be the differences in their appearances and properties.

The conjecture has naturally long been hazarded that these different atoms might be only differently arranged aggregates of a simple fundamental atom and that the lightest, the atom of hydrogen. As early as 1815 Prout proposed the hypothesis that all atoms are derived from the hydrogen atom. Now, in that case, the atomic weights must be exact multiples of the atomic weight of hydrogen, or, since that is unity, must be whole numbers. To settle this question, very accurate atomic weight determinations are necessary. If, for example, the atomic weight of mercury is found in a first fairly accurate determination to be 200, this value is in agreement with Prout's hypothesis. But if a more accurate measurement gives 200·6, this is in contradiction to Prout's hypothesis. As a matter of fact, only a few atomic weights are exact integers, but in very many cases the deviation from a whole number is only at the most $\pm 0 \cdot 1$.

If, then, no explanation can be found for these deviations of the atomic weights from whole numbers, Prout's hypothesis in this simple form cannot be correct. Nevertheless, out of the sixty-three elements whose atomic weights are fairly accurately known, the number of cases in which the atomic weights vary by not more than $\pm 0 \cdot 1$ from a whole number is 27, while it should be only 18 or 19 on the theory of probability, so that there is certainly some grain of truth in Prout's hypothesis, at any rate for a large number of atoms.

On the other hand, it has long been noted that the atoms of substances which are chemically similar show a number of approximate regularities in their atomic weights in that the successive substances exhibit practically constant differences of atomic weight. These approximate regularities have found

THE PERIODIC SYSTEM OF THE ELEMENTS.

O.	I. a	I. b	II. a	II. b	III. a	III. b	IV. a	IV. b	V. a	V. b	VI. a	VI. b	VII. a	VII. b	VIII.
Helium He 4·0	Lithium Li 6·94		Beryllium Be 9·1		Boron B 14·0		Carbon C 12·0		Nitrogen N 14·0		Oxygen O 16·0		Fluorine F 19·0		—
Neon Ne 20·2	Sodium Na 23·0		Magnesium Mg 24·3		Aluminium Al 27·1		Silicon Si 28·3		Phosphorus P 31·0		Sulphur S 32·0		Chlorine Cl 35·5		—
Argon A 39·9	Potassium K 39·0		Calcium Ca 40·1		Scandium Sc 45·1		Titanium Ti 48·3		Vanadium V 51·0		Chromium Cr 52·0		Manganese Mn 54·9		Iron, Cobalt, Nickel Fe 55·8, Co 58·9, Ni 58·7
		Copper Cu 63·5		Zinc Zn 65·4		Gallium G 69·9		Germanium Ge 72·5		Arsenic As 74·9		Selenium Se 79·2		Bromine Br 79·9	—
Krypton Kr 82·9	Rubidium Rb 85·4		Strontium Sr 87·6		Yttrium Y 88·7		Zirconium Zr 90·6		Niobium Nb 93·5		Molybdenum Mo 96·0		—		Ruthenium, Rhodium, Palladium Ru 101·7, Rh 102·9, Pd 106·7
		Silver Ag 107·9		Cadmium Cd 112·4		Indium In 114·8		Tin Sn 118·7		Antimony Sb 120·2		Tellurium Te 127·5		Iodine I 126·9	—
Xenon X 130·2	Caesium Cs 132·8		Barium Ba 137·3		Lanthanum La 139·0		Cerium Ce 140·3		—		—		—		—
—									Tantalum Ta 181·5		Tungsten W 184·0		—		Osmium, Iridium, Platinum Os 190·9, Ir 193·1, Pt 195·2
—		Gold Au 197·2		Mercury Hg 200·6		Thallium Tl 204·0		Lead Pb 207·2		Bismuth Bi 209·0		—		—	—
—			Radium Ra 226·0		—		Thorium Th 232·1		—		Uranium U 238·2		—		—

[To face page 7.

their most general expression in the *periodic system of the elements*, drawn up by Mendelejeff and by Lothar Meyer (1869), which states that the chemical properties of the atoms are periodic functions of their atomic weights. For a large number of chemical elements this law, as the attached table shows, is perfectly fulfilled, for when the elements are arranged in the order of their atomic weights, the ninth element has the same chemical properties as the first. But here, again, there is no universal regularity. For the atoms up to manganese (atomic weight 54·9) the regularity is good. But then come the three elements of the iron group, iron, nickel, and cobalt, which do not fit into this series and have very slightly different atomic weights. Then the elements fit again fairly well until the palladium group (ruthenium, rhodium, and palladium) brings out a fresh discordance. The same thing happens a third time with the platinum group, osmium, iridium, and platinum. The difficulty became greater when a very large number of new elements were discovered among the rare earths. And further, it appears that elements which belong to one column and so should be chemically similar to one another may be sub-divided into two groups, distinguished by marked differences. In each column, therefore, two series, *a* and *b*, must be distinguished. It is hence impossible to fix the order in the periodic system without a certain amount of arbitrariness, and it is, in fact, often doubtful, especially for the elements in the middle, into what columns they should be arranged. But we now know from the investigations on X-ray spectra, which will be considered later, firstly, what is the right serial order of the elements—and it does not always agree with the order of the atomic weights—and, secondly, where there are still gaps in the system, still unknown elements. Many different arrangements have been proposed for the periodic system. In the subjoined table is given one of them which does not include the new rare earths ; in it an extra column, O, has been inserted before the columns I-VII, drawn up by Mendelejeff and Lothar Meyer, which accommodates the recently-discovered rare gases, and there is also a column, VIII, in which the above-mentioned elements, which do not fit into the series, are included. This arrangement is incomplete, and some of the elements of the rare earths are missing from the table. They come between cerium and tantalum. Hydrogen, which occupies a unique position, is not included. In the fourth lecture a more extended and complete table will be given.

It will be recognized from an inspection of the elements which come underneath one another that in order to show similarities, the separate columns must be sub-divided into two groups, *a* and *b*, such that the elements in *a* or in *b* are very similar. Thus, in column I, lithium, sodium, potassium, and rubidium are substances which belong together, and copper and silver and gold go together, but the former and the latter have no direct chemical resemblance. The same sort of thing happens in the other columns.

The periodic system may certainly be claimed to give an indication, a first approximation, to the true connection between the different atoms, but there are many points about it which are still obscure.

For a more accurate conception of the atoms, it is most important to obtain at this stage definite figures, if possible, for their size and mass. The atomic weights give only the relative masses of the individual atoms, taking that of hydrogen as unity. It is of particular interest to inquire what is the actual mass of an atom in grammes, and what is the radius of an atom assuming it to be spherical. Direct measurement is impossible, for the atoms are so small that our most powerful microscopes can never show us anything of them, and their mass is so minute that our balances would not enable us to determine it even if they could be made a million times more sensitive than they are.

There are, however, indirect methods by which we may obtain fairly accurate ideas of these dimensions. In the first place, it is true, these methods give us only information about the molecules, but as we know that the separate molecules are composed of a known number of atoms we know from this what we want about the atoms. One of these indirect methods depends on the *Kinetic Theory of Gases*. A body may exist either in the solid, liquid, or gaseous state. It is obvious that the molecules of solid and liquid bodies must be much closer together, that they must fill space much more completely than they do in the gaseous state. For the observed density or specific gravity of a body, i.e. the ratio of the mass of a body to its volume, is much greater in the solid and liquid states than in the gaseous. Since the molecules in the gaseous state are, on the average, at relatively great distances from one another we may further assume, as the Kinetic Theory of Gases does, that the forces which, in general, exist between molecules and which decrease in magnitude with increasing separation of the molecules, become vanishingly small for substances in the gaseous state.

According to this theory the properties of gases depend on the fact that the individual molecules of a gas are not at rest but in lively motion.   Now since the forces between the molecules disappear on account of the relatively great separation, that is, there is no force acting on a molecule, its motion can only be in a straight line with constant velocity. Every molecule, then, travels about in straight lines.   Since there are a large number of molecules even in a small space, the molecule must soon collide with another.   In this proximity to a second molecule, molecular forces will be operative.   Their effect may be interpreted as being that a molecule which meets another, collides with, and rebounds from it, like a billiard ball which strikes another. The same occurs when a molecule hits the wall of the vessel.   We must, therefore, conceive of a gas as consisting of an enormous number of molecules which continually dart about in confusion and are always colliding with other molecules, and consequently changing the direction and magnitude of their velocity : between any two collisions each molecule moves in a straight line with constant velocity.   Every individual molecule thus describes a zig-zag path, consisting simply of straight sections of greater or less length.

This assumption gives an explanation of all the properties of gases. In the first place, the pressure exerted by the gas on the walls of the vessel arises from the impacts of its molecules on the wall. The more molecules there are in a given space, that is, the greater the density of the gas, the greater is the pressure. As the density of a given mass of gas is inversely proportional to its volume, this holds also for the pressure. The velocity of each separate molecule varies continually, and thus all the molecules which at a given moment are contained in a space, say of 1 c.c., have very different velocities. The mean of the simultaneous velocities of all the molecules is a characteristic for the behaviour of the gas, and as the forces exerted in collision, and hence the pressure of the gas, depend both on the mass of the molecule and on this *mean velocity*, observation of the pressure allows us to calculate the mean velocity. The latter is naturally found to be different for different gases, the lighter gases having greater, and the heavier smaller, velocities. The mean velocity is independent of the pressure of the gas, but increases with increase of temperature, actually as the square root of the absolute temperature. (The absolute temperature $= 273°$ + the ordinary centigrade temperature : a gas at $17°$ C. has an absolute temperature of $290°$.) The mean velocity of a molecule is very high, amounting to several hundred metres per second. At $0°$ C., for example, it is 447 m./sec. for air, 1692 m./sec. for hydrogen, 362 m./sec. for carbon dioxide.

Notwithstanding these high velocities, however, the molecules remain nearly in the same place because they very frequently collide with other molecules, rebound, collide again, and so on. The path of a molecule between two consecutive collisions is sometimes longer, sometimes shorter, and so is very different for different molecules at the same time. But here, again, we may take the mean of all these paths, which we call the *mean free path*, and this magnitude determines a number of phenomena which gases offer for observation, in particular the viscosity and thermal conductivity of gases. The thermal conductivity, for example, which causes the spontaneous equalization of initially different temperatures between two parts of a gas can be explained immediately by the Kinetic Theory. For in the warmer part the molecules have on the average a greater velocity, and hence greater kinetic energy, than in the colder part. The more rapid molecules are continually darting across the boundary between the two parts of the gas, from the warmer part to the colder, and slower molecules from the colder to the warmer. Thus, the content of energy, i.e. the temperature of the warmer part decreases, and that of the colder increases until both parts are at the same temperature. Equilibrium will be the more rapidly attained (apart from the differences in velocity) the greater the mean free path of the molecule is, i.e. the farther a warmer molecule can penetrate into the colder part. It follows, inversely, that the mean free path may be determined from observation of the conductivity (and similarly of the viscosity). The values found are, as was to be expected, very small. They depend on the density, and hence on the pressure of the gas : the more rarefied the gas is the greater, of course, is the mean free path, simply because the molecules are then on the average farther from each other. It is found that, e.g. at atmospheric pressure, the mean free path for air $= 0.96 \times 10^{-5}$ cms. ; for hydrogen $= 1.78 \times 10^{-5}$ cms. ; for oxygen, 1.02×10^{-5} cms. ; for nitrogen, 0.95×10^{-5} cms. ; for carbon dioxide, 0.65×10^{-5} cms.

These two magnitudes, the mean velocity and the mean free path, have been introduced in order to find the answer to the question which we propounded, viz. : " What is the actual number of molecules in a cubic centimetre of a gas at atmospheric pressure ? " The number we are seeking is called the *Loschmidt number*, because Loschmidt first gave a method for its determination. The second question which particularly interests us, namely—" How great is the radius of a molecule if we assume it spherical ? "—is closely related to this. For if we know this radius we know also the volume of a molecule, and if we know also the Loschmidt number we know the total volume actually occupied by the molecules contained in 1 c.c. of the gas. Now this volume will be approximately the same as that which these molecules occupy if the gas is converted by pressure or by cooling into a liquid or into a solid. For then we may assume that the molecules will be packed together as closely as possible. (Actually the liquid will still have a greater volume than corresponds with the sum of the volumes of all the molecules.) For this reason the density of a substance in the liquid or solid state is far greater than in the gaseous.

Gaseous nitrogen, e.g. at atmospheric pressure, has a density of 0·001254, while the greatest density observed for liquid nitrogen is 0·854 : the ratio of the volume of the molecules when in close contact to the volume of the same molecules in the state of gas at atmospheric pressure is, for nitrogen,

$$\frac{0·001254}{0·854} = 0·00138.$$

This number is known as the *coefficient of condensation*. Now if we reflect that the centre of a molecule in the closely packed condition must be displaced by an amount equal to the radius before it reaches a second molecule, while in the gaseous state the centre of a molecule must be displaced by the mean free path before it reaches a second molecule, we realize without any accurate calculation that the coefficient of condensation will be roughly equal to the ratio of the radius of a molecule to the mean free path. Hence the radius of a molecule may be calculated from the mean free path and the coefficient of condensation. For nitrogen, e.g. which has a coefficient of condensation of 0·00138 and a mean free path of $0·95 \times 10^{-5}$ cms., the radius comes out on this line of reasoning to about $0·00138 \times 0·95 \times 10^{-5} = 1·30 \times 10^{-8}$ cms. We are not concerned at the moment with the exact figure, but only with the order of magnitude 10^{-8} cms. More accurate calculation by this method has given the following values for the radii of the molecules of various substances :—

				Radius of a Molecule.
Hydrogen	.	.	.	$2{\cdot}56 \times 10^{-8}$ cms.
Helium .	.	.	.	$2{\cdot}20$,,
Nitrogen .	.	.	.	$3{\cdot}51$,,
Oxygen .	.	.	.	$3{\cdot}38$,,
Argon	.	.	.	$3{\cdot}41$,,
Chlorine .	.	.	.	$5{\cdot}04$,,
Mercury .	.	.	.	$5{\cdot}86$,,
Ether	.	.	.	$7{\cdot}30$,,

The radius of a molecule is not to be taken here in a strict sense. Round each molecule there must exist a sphere of influence such that another molecule cannot penetrate into it. The numbers given above are really the radii of the spheres of influence rather than the radii of the molecules themselves i.e. they are too high. It is only for the monatomic molecules (helium, argon, and mercury in our table that the numbers obtained give also the radii of the atoms. For the diatomic molecules, such as those of hydrogen, oxygen, nitrogen, and chlorine we can only say that their atoms, if they are spherical must have smaller radii than indicated by the num bers given. The result of these applications of the Kinetic Theory of Gases gives us, then, a clue to the actual sizes of the atoms or rather to their order of magnitude. *The radius of an atom (supposed spherical) must be assumed to be of the order of* 10^{-8} *cms.* Different atoms have radii rather greater or less than 10^{-8} cms. The same magnitude is also indicated by other quite independent considerations with which, however, we shall not at present concern ourselves. The magnitude 10^{-8} cms. is the 10 tmillionth part of a millimetre.

The determination of the size of a molecule renders it easy to obtain a definite value also for the number of molecules in a cubic centimetre, i.e. for the Loschmidt number. For, if we consider that on the average every molecule traverses the mean free path before it encounters another molecule, we see that the cylindrical space which a molecule describes in this flight is of such a size that there is only 1 molecule in it on the average. The ratio of 1 c.c. to this space is equal to the number of molecules in 1 c.c., i.e., to the Loschmidt number.

Calculation shows that at atmospheric pressure there are in 1 c.c. of gas 27·2 trillion molecules. A trillion is 10^{18}. This Loschmidt number has the same value for all gases, for according to Avogadro's Hypothesis, all gases at the same temperature and pressure contain the same number of molecules per cubic centimetre.

If we assume these molecules uniformly distributed in the cubic centimetre, each molecule has a small cube of content

$$\frac{1}{27\cdot 2 \times 10^{18}} \text{ c.c.}$$

to itself.

The length of the side of this cube is, therefore, about

$$\frac{1}{3 \times 10^{6}} \text{ cms.} = 3\cdot 3 \times 10^{-7} \text{ cms.,}$$

i.e. 3 or 4 millionths of a millimetre. This, then, is the value of the mean distance between 2 molecules. The distance between 2 molecules is thus on the average some 10 to 30 times as great as the radius of a molecule.

From the Loschmidt number another important number may be calculated which is known as the *Avogadro number*. If we take 1 c.c. of different gases at the same pressure, say hydrogen, oxygen, and chlorine, the masses of these equal volumes are not the same, but are in the ratio of the molecular weights, viz. in this case as $2 : 32 : 71$. In general, equal volumes of different gases have masses which are in the ratio of their molecular weights, hence it follows that the volumes which are occupied by 1 grm. molecule of any gas at the same temperature and pressure are equal and so, according to Avogadro, contain the same number of molecules. This number, the number of molecules contained in a gramme molecule of any gas at a pressure of one atmosphere and at $0°$ C., is known as the Avogadro number. It may readily be determined by calculating how many cubic centimetres are occupied by a gramme molecule (2 grms.) of hydrogen. A gramme of air at N.T.P. occupies a volume of 773 c.c., and hydrogen is 14·475 times as light as air ; hence 2 grms. of hydrogen (a gramme molecule) occupies a space of $2 \times 14·475 \times 773 = 22,378$ c.c. Further, a gramme molecule of hydrogen, and consequently a gramme molecule of any other substance, in the gaseous state at N.T.P. contains

$$27·2 \times 10^{18} \times 22,378 = 60·9 \times 10^{22}$$

molecules. This, then, is the Avogadro number.

From it we can now calculate the actual mass of a molecule of hydrogen and thence that of any other molecule. For since $60·9 \times 10^{22}$ molecules

2

of hydrogen together possess a mass of 2 grms., each molecule of hydrogen must have a mass of

$$\frac{2}{60 \cdot 9 \times 10^{22}} = 3 \cdot 28 \times 10^{-24} \text{ grms.}$$

10^{-24} grms. is 1 grm. divided by a quadrillion. A molecule of oxygen has, of course, a mass 16 times, a molecule of chlorine $35\frac{1}{2}$ times as great. An atom of hydrogen accordingly has half the above mass, viz. $1 \cdot 64 \times 10^{-24}$ grms. This number will serve later for a further calculation.

All these numbers for the size, mass, and number of molecules and atoms tell us nothing directly that we can picture to ourselves at all. The ratio of the unit with which we are accustomed to reckon to these minute or enormous values is too different. For a unit of length we are accustomed to use 1 mm., and we cannot conceive of the 10 millionth part of it, the radius of an atom. We can get a fair idea of it if we magnify our unit (1 mm.) 10 million times so that it becomes 10 kms. ; then the radius of an atom would correspond with a length of 1 mm.

It is still more difficult to compare the mass of an atom with our usual units. Here we must use the earth to help us. The whole earth has a mass of 2×10^{24} kg. The mass of a hydrogen atom bears about the same ratio to the mass of a gramme as the mass of 3 kg. to that of the earth.

And the most recent investigations, which we shall consider in these lectures, show that these atoms, these minutest particles of matter, are themselves regular worlds, that each of them represents a veritable solar system.

The simplest assumption to make about the atoms according to the facts we have hitherto considered is, that they are very small, spherical or otherwise shaped masses of the respective chemical substances. A molecule would correspond with a juxtaposition of two or more atoms of the same or of different substances. According to this primitive assumption, there are shown in Fig. 1 in the upper row a hydrogen atom and a chlorine atom, while the lower row gives a representation of a hydrogen molecule and of

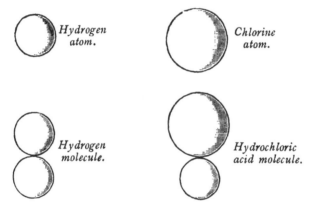

FIG. 1.—Hydrogen atom; chlorine atom; hydrogen molecule; hydrochloric acid molecule.

a hydrochloric acid molecule.   This assumption gives no explanation of why two hydrogen atoms always unite more closely to form a hydrogen molecule.

The atoms of different substances are, as their name (ἀ-τομός, indivisible) implies, assumed to be absolutely unchangeable, indivisible particles. But now there are two different possibilities. One would be that, though the atoms appear to us indivisible, it is only because we are not yet able to divide them, because we have not, or not yet, the power to do so. An atom of gold, as such, would be indivisible.

But since it occupies space, we might think of it as composed of still smaller parts, only these parts would no longer be gold, but something else of which gold itself was composed. This view would, therefore, consider the indivisibility of atoms as something provisional or relative, and it is obviously related to Prout's hypothesis above mentioned, that the atoms themselves are made up of other constituents, except that these constituents need not be precisely the hydrogen atom. This idea has been shown by the investigations of the last decade to be well founded.

Let us develop the other assumption, however, namely, that for some reason or other atoms are absolutely indivisible quantities. In that case, indivisibility must be fundamental to their nature, and not merely due to the defectiveness of our resources. And, obviously, inquiry as to how this could be possible would be legitimate and interesting, even if we could decompose the atoms themselves into smaller particles or proto-atoms. For, then, the question would still remain, Why are these proto-atoms really indivisible? The question is, then, how could the indivisibility of a certain quantity of matter be explained? A satisfactory answer is afforded by a very ingenious theory which was propounded fifty years ago by the great English physicist, Lord Kelvin, the *theory of the vortex atom*. It was shown by Helmholtz that vortex rings could exist in a non-viscous fluid. They are ring-shaped structures in which the particles of fluid are conceived as rotating round the middle line of the ring as illustrated in Fig. 2. It follows, however, from the general properties of such a fluid that a vortex of this kind cannot be produced by the ordinary simple, so-called conservative, forces of nature, nor be destroyed by them. Such vortices must rather, if they ever exist in a fluid, have existed in it from the beginning of creation, and they cannot be disrupted or destroyed by conservative forces. If you cut at such a vortex ring with the sharpest knife imaginable and try to sever it, you fail. The vortex ring changes its shape, it bends round the knife edge, but the whorling particles cannot be separated, the fluid of the vortex preserves its coherence unimpaired. Vortex rings and their properties may be dimly recognized in the smoke rings which accomplished smokers blow in the air. Only dimly, for air is not a fluid of no viscosity, such as is postulated in Helmholtz's statements. As a matter of fact,

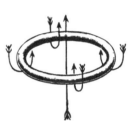

Fig. 2.

these smoke rings may be seen to dissolve after lasting for a time, but that is only because there are here viscous forces, forces which are not conservative, present in the fluid. In a fluid devoid of viscosity, the parts in vortex motion hold together for ever, and form a ring which is always closed, even though its shape may change. Such a vortex ring, then, is in truth something indivisible, an ἄ-τομός. Now Kelvin assumes that the ether, which we must suppose omnipresent, all-pervading— for it is the medium for the motion of light, and light is all-pervading—that this ether has the properties of a non-viscous fluid. Suppose that in it many such vortices have existed from the beginning. These vortices, which are indestructible, form the atoms of bodies and the atoms are *vortex atoms*. Since we cannot produce them with our forces, the number of atoms in the world cannot be increased. Neither can it be diminished, for we cannot destroy such vortices. In short, these vortices have the essential properties of atoms, indestructibility and indivisibility. Although, as we have remarked, the indivisibility of atoms can now no longer be assumed to be a fact, nevertheless Kelvin's theory is still of great philosophical value, for the same question as to the possibility of indivisibility has merely been transferred from the atoms to the proto-atoms. At the present time, when positive and negative electrons are recognized as the constituents of atoms, this theory might conveniently be designated the vortex theory of electrons.

For a long time a number of difficulties have stood in the way of assuming that atoms were simple, indivisible particles. In the first place, optical phenomena have to be considered. As is well known, all luminous gases and vapours show very characteristic phenomena when the light from them is passed through a slit and a prism or diffraction grating. Whereas sunlight or the light from a white-hot arc lamp gives, when examined in this manner, a complete spectrum, a band of colour which extends from red to violet, and exhibits every transition from red through orange, yellow, green, bright blue, dark blue, to violet, the case is quite different with coloured gases and flames. When a sodium salt, common salt, or Glauber's salt, is introduced into the flame of a Bunsen burner, the flame is coloured bright yellow; if we pass this light through a slit and a prism, we obtain instead of a spectrum only a sharp yellow line, known as the D line; with a prism of sufficient dispersion, this is recognized as a double line, D_1 and D_2.

Since the same line is given by all salts of sodium, the chloride, the sulphate, the nitrate, etc., it must have its origin only in the atom of sodium. Similarly, a flame coloured red by lithium gives two lines in the spectrum, a red and an orange. A potassium flame gives a red and a violet, a cæsium flame one line in the red, one in the orange, one green, and two blue lines. As light consists of vibrations, and each colour corresponds with a definite vibration frequency, it follows that an atom of a body may, according to circumstances, execute vibrations of different frequencies. Even the simple gases may, if they are in rarefied form, be made luminous by sending through them an electric current, as in the Geissler spectrum tubes. Thus hydrogen, for example, shows, to begin with, four lines, a red, a green, a blue, and a violet. With helium seven lines are visible, and argon, krypton, and xenon give spectra with numerous lines. In the spectrum of the vapour of iron more than two thousand lines have been found. These optical vibrations, in so far as we have been able to apply rules to them, are governed by laws quite different from those for the vibrations of stretched bodies. It follows that vibrations or periodic movements of very different kinds must be occurring in the atoms, which give rise to this emission of light. Accordingly, the conclusion may be drawn from the multiplicity of lines in the spectra of elements, that the atoms cannot be simple bodies, but must contain parts of some sort, i.e. that they cannot really be " atoms." These observations, which have been known for years, teach us that there are still important problems about the atoms, which physics and chemistry have long accepted as the ultimate units of matter, and that we now have to unravel the constitution of these apparently simple bodies.

LECTURE II

ATOMS AND IONS AND ELECTRICAL EFFECTS IN
LIQUIDS AND GASES—ATOMS OF ELECTRICITY

T HE advances in our knowledge of the nature
of atoms have come from the study of electricity. In the middle of the last century
new discoveries were continually being made in the
realm of electricity ; the heating effects, the elec-
trolytic effects, the magnetic and electro-dynamic
effects of the electric current were investigated.
When Faraday discovered the induction effects,
and Hertz the propagation of them by electric
waves, it became increasingly difficult to form a
consistent idea of the nature of electricity. The
diversity of the effects was so great that it seemed
almost impossible to include them all under one
head. Many of the effects of electricity were such
as to indicate that electricity was a substance,
others seemed to demand the assumption that
electricity depended on some disturbance of the
ether. In fact, physicists frequently changed their
views on the subject, and when Hertz had made
his splendid discoveries about electric waves and
radiation, the ether theory, the motion theory of
electricity, seemed to have clearly won the day.
The theory known as Maxwell's, which explained all
electrical phenomena as static or dynamic effects

of the ether, seemed at that time to be indubitably the most complete theory of electricity.

So it seemed.   But on closer examination a number of important phenomena remained unexplained even by its aid : these were precisely those which could, on the contrary, be very simply explained by assuming a corpuscular nature for electricity. The phenomena of electrolysis in particular, investigated and reduced to rule by Faraday, could not be explained on Maxwell's theory, but seemed to point instead to electricity as being corpuscular.

It is well-known that, when an electric current is passed through a compound liquid conductor (a solution of salts, acids, or bases in water), the latter is decomposed or electrolyzed in such manner that the metal (or hydrogen) of the dissolved salt, acid or base, is liberated at the negative pole or electrode, the *cathode*, and the remainder · of the molecule at the positive electrode, the *anode*.   The two constituents of the molecule which are thus separated, showing themselves to be electrically charged constituents, are called its *ions*, that which appears at the negative electrode being known as the *cation*, and that at the positive as the *anion*.   Investigation of the exact quantitative phenomena of electrolysis led *Faraday* to the law, that the weight or mass of an ion, which is separated in the electrolysis of any liquid at either electrode, is proportional to the product of the equivalent weight of the substance in question, and the total amount of electricity which has flowed through the liquid during the passage of the current.

The mass of the separated ion is proportional to

equivalent weight of the ion × quantity of electricity.

The equivalent weights of elementary bodies are their atomic weights divided by their valencies. Hydrogen has an equivalent weight 1, divalent oxygen an equivalent $\frac{16}{2} = 8$, trivalent nitrogen an equivalent $\frac{14}{3} = 4.66$, tetravalent carbon an equivalent $\frac{12}{4} = 3$.

The total quantity of ions separated, divided by their equivalent weight, gives simply the number of equivalents separated. If we express the total quantity separated in grammes, we obtain the number of gramme-equivalents separated ; a gramme-equivalent of a substance is that number of grammes which is indicated by the equivalent weight. If 10 grms. of oxygen are separated in an electrolysis, that is $\frac{5}{4}$ gramme-equivalents of oxygen. Faraday's law may, therefore, be expressed also thus : the number of equivalents separated is proportional to the quantity of electricity which has flowed ; or inversely, the quantity of electricity which has flowed is proportional to the number of equivalents separated.

Many experiments show that the conduction of a current through a liquid may be interpreted as follows : there are originally a very large number of ions of both kinds free in the liquid, an equal number of positive and negative ions occurring in any small space. If two electrodes are now dipped into the liquid and connected with the source of current, electrostatic attraction causes the positive ions to migrate in the direction of the cathode and the negative ions in the direction of the anode. Those positive or negative ions which happen to be nearest to the cathode or anode separate there, while in the interior of the liquid there remain always as many positive as negative ions. But this process is an indication that every ion which separates was previously taking part in the conduction of the current. Conduction through the liquid takes place always by *migration of the ions*, and these ions are simply positively or negatively charged parts of molecules. Every equivalent of a substance which separates was previously, therefore, a migrating equivalent, and carried a certain quantity of electricity. Now, Faraday's law mentions only the number and not the kind of equivalents ; so it must predicate that each equivalent of any substance is combined with one and the same quantity of electricity, with a positive quantity if it is a cation or with a negative if it is an anion. Every monovalent atom, such as those of hydrogen, silver, chlorine, etc., is accordingly combined with a definite quantity of electricity, every divalent atom, such as those of copper, zinc, or oxygen, with twice as much, every tri- or tetra-valent with three or four times as much.

Helmholtz (1881) was the first to draw a most important conclusion from the above facts. The simplest interpretation of the law just stated is that electricity, like matter, is divided into atoms, and that an absolutely definite quantity of electricity is combined with each equivalent of any substance, this quantity being the same no matter what the substance is. There is a certain minimum quantity of electricity which is combined with each atom of a monovalent substance. A substance of higher valency cannot contain, say, $1\frac{1}{2}$ or $3\frac{1}{2}$ times this minimum quantity per atom, but only 2, 3, etc., times as much. Electricity seems, therefore, to be a substance, and one, too, which is divided into very small discrete parts or atoms. These atoms of electricity we call *electrons* and, to begin with, we distinguish positive and negative electrons. Electricity behaves like a special chemical element, whose atoms combine with those of the other known elements to form ions. And it is a monovalent substance; for a monovalent atom combines with one electron, a bivalent with two, and so on, exactly as a chlorine atom combines with one atom of hydrogen, and an oxygen atom with two atoms of hydrogen.

Such was the definite conception which was first introduced into physics by Helmholtz, that electricity was a material divided up into atoms, a conception which was in flat contradiction to Maxwell's theory. On the other hand, a very large number of electrical phenomena could be simply and completely explained according to Maxwell's theory by disturbances of the ether, so that here again it seemed impossible to bridge the gulf that yawned between the different types of manifestation of electricity. This contradiction was, however, reconciled and a compound synthesized out of the different conceptions by the *electron theory of electricity*, propounded by H. A. Lorentz (1892). Lorentz pointed out that all the effects of electricity inside bodies could be explained on the atomistic theory by the assumption of electrons, and that, *per contra*, all the effects of electricity at a distance, the electrostatic, electromagnetic, and inductive effects, required the help of the ether.   He was able to unite these two classes of phenomena by showing that every electron is closely bound up with the ether, that an electron at rest and one in motion each produced quite definite changes in the ether, which were then propagated through the latter—with the velocity of light—and so caused actions at a distance. We need not for our purpose go further into the development of this electron theory ; but what is specially important for us is that we must now include in our consideration two more atoms besides those of the known chemical elements, namely, atoms of positive and negative electricity.   True, this assumption has so far been deduced only from complicated and rather involved facts, and the supposed atoms of electricity appear to occur only in combination with atoms of ordinary substances. But we shall now discuss other phenomena, which will lead us to recognize that these electrons may, under certain circumstances, exist in the free state, uncombined with other matter, so that their existence is no longer to be deduced only from a difficult interpretation of complex phenomena.

In order to fix these ideas completely and to lay at the same time a foundation for later generalizations, we must now discuss the consequences resulting from Faraday's law, not merely qualitatively, but quantitatively, by numerical measurement.

We stated the law—" the quantity of electricity which passes is proportional to the number of equivalents separated "—and expressed it also in another way—" each equivalent weight of a substance is combined with a definite quantity of electricity, independent of the nature of the substance."

As the equivalents of the elements are, in the first place, relative numbers compared with the equivalent weight of hydrogen taken as unity, we may ask our first quantitative question thus—" What is the quantity of electricity which is combined with one gramme equivalent of any substance ? "

Before answering this question, we must first decide in what units to measure quantities of electricity. In electrical practice, the unit of quantity of electricity is that which is carried per second by a current of 1 ampere across each section of a circuit. This unit quantity of electricity is known as a *coulomb*. For many scientific calculations it is, however, more convenient to take for the unit quantity of electricity that which exerts on an equal quantity placed at a distance of 1 cm. from it an electrostatic force of attraction or repulsion which is equal to the unit of force, 1 dyne. This latter unit has not been given any simple special name, as it is not much used in practice ; we call it the *electrostatic unit of quantity of electricity*. It is much smaller than a coulomb, and accurate measurements have shown that a coulomb contains three thousand million (3×10^9) electrostatic units. Given this, a quantity of electricity in coulombs may be converted to electrostatic units, or *vice versa*.

First let quantity of electricity be measured in

coulombs. Then the quantitative question above propounded is answered by the following experiment. This was carried out by Lord Rayleigh and Miss Sidgwick in England (1884) and by F. and W. Kohlrausch in Germany (1886) with extreme precision and with exactly the same result, and forms the basis of the international standardization of the unit of current, the ampere. If a quantity of electricity equal to 1 coulomb is passed through a solution of silver nitrate, 0·001118 grms. of silver are deposited at the cathode. On the above explanation this means that 0·001118 grms. of silver are combined with 1 coulomb; consequently a gramme-equivalent of silver (107·88 grms.) is combined with

$$\frac{107·88}{0·001118} = 96,494 \text{ coulombs.}$$

According to Faraday's law, the same number of coulombs is combined with 1 grm.-equivalent of any other substance, e.g. with 1 grm. of hydrogen or 35·5 grms. of chlorine, with 8 grms. of oxygen or 31·78 grms. of bivalent copper, or with 9·03 grms. of trivalent aluminium.

This shows at once that very different quantities of electricity are combined with a gramme of different substances, being the smaller as the equivalent weight becomes greater. The quantity of electricity combined with 1 grm. of a substance is called the *specific charge* of that substance; this is equal to the ratio of the charge (quantity of electricity) to the mass of that substance. From the above numbers it appears that the specific charge of hydrogen, the element with the smallest equivalent weight = 96,494 coulombs per gramme; that of oxygen is $\frac{96494}{8}$ = 12,062 coulombs/gramme, that of chlorine $\frac{96494}{35\cdot5}$ = 2718 coulombs/gramme, that of silver $\frac{96494}{107\cdot88}$ = 894·4 coulombs/gramme and so on.

The specific charge of any mass in electrolysis is equal, at the most, to 96,494 coulombs/gramme, which is its value for hydrogen. For all other elements it is less than 96,494 coulombs/gramme, and is the smaller the greater the equivalent weight of the element concerned.

We shall draw an important conclusion from this law later on. The calculation we have just made enables us to make a further great advance. Since every monovalent ion has a certain charge, namely, the charge of one electron, we can now fix the magnitude of this charge. A gramme equivalent of hydrogen, i.e. 1 grm., is combined with 96,494 coulombs. Now, a gramme molecule of hydrogen i.e. 2 grms., contains (see p. 17) 60·9 × 10²² molecules (the Avogadro number), so that 1 grm. of hydrogen contains the same number of atoms or ions. Therefore, each ion of hydrogen (and consequently each ion of any other monovalent substance) contains a charge of

$$\frac{96494}{60\cdot9 \times 10^{22}} = 1\cdot58 \times 10^{-19} \text{ coulombs.}$$

We prefer to express this number in electrostatic units, by multiplying by 3 × 10⁹, which gives

$$4\cdot74 \times 10^{-10} \text{ electrostatic units.}$$

This number is known as the *elementary electric charge* or the *elementary quantum of electricity*.[1] It is obviously the charge of a single electron, and is thus the smallest electrical charge which ever exists. A charged body can contain only a whole number of these quanta of electricity, not a whole number plus a fraction of them. But as this elementary quantum is so small that it cannot be detected with our most refined instruments, this conclusion cannot be directly proved.

If we visualize the above conceptions, we shall picture a positive or negative ion somewhat as in

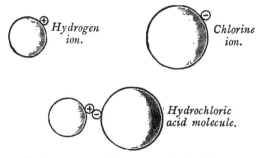

FIG. 3.—Hydrogen ion, chlorine ion, hydrochloric acid molecule.

Fig. 3. The electrons are represented by small circles distinguished by + and − ; a hydrogen ion consists of a combination of a hydrogen atom with a positive electron, and a chlorine ion of a combination of a chlorine atom with a negative electron. The combination of chlorine and hydrogen to form a hydrochloric acid molecule can be understood, as in Fig. 3, to be the union of a positive and negative ion. But the combination of two hydrogen atoms to form a hydrogen molecule is incomprehensible on

---

[1] This is more commonly designated in England as the "ionic charge," abbreviated as "e."—Tr.

this view, for each ion of hydrogen in electrolysis is always positively charged, and two positive charges do not attract, but repel, one another.

Although the atomistic structure of electricity was first revealed by the phenomena of electrolysis, it should be mentioned that the phenomena are quite complicated, and that one cannot be quite sure that this explanation, though relatively simple, is really valid and will stand further test.

An essentially new insight into our electron hypothesis, together with a verification and extension of it, was obtained by observation of the passage of electricity through very rarefied gases. Let two metal plates or wires be sealed as electrodes into a glass vessel, and let the gas pressure in the vessel be reduced by a vacuum pump to some 0·01 mm. or less. If the electrodes are connected with a high tension source of electricity—an induction coil, a Wimshurst machine or a high tension battery—very striking and characteristic phenomena, which are called *cathode rays*, appear in the vessel. The interior of the vessel, at this degree of exhaustion, remains practically non-luminous, but it is seen that the glass of the tube exhibits a green fluorescence, this being strongest in the parts immediately opposite the cathode or negative electrode. The cause of this fluorescence, as the following experiments will show, is that invisible rays proceed in straight lines from the cathode, and wherever they meet an obstacle, such as the glass of the tube, produce peculiar effects ; in the present case they excite fluorescence. It may readily be shown that these rays travel only in straight lines—that is why we call them rays—do not bend round corners, and proceed only from the

FIG. 4.

cathode. For this purpose a V-shaped tube like Fig. 4 is used. If in this we make L the cathode and M the anode, only the left arm of the glass up to the bend is seen to glow green, not the right. *Vice versa*, if M is made the cathode and L the anode, the right arm fluoresces. So the cathode rays do not travel from the cathode to the anode past the bend in the tube, but only in straight lines from the cathode. That their propagation is in absolutely straight lines is very clearly shown, too, by a

tube like that in Fig. 5. Inside this tube a cross, *b*, of metal, is fixed opposite the cathode, *a*. Cathode

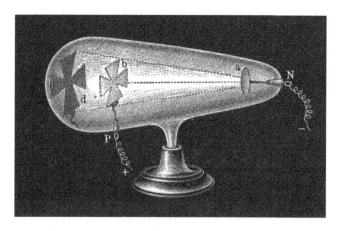

FIG. 5.

rays are emitted in straight lines from *a*, and excite a green fluorescence where they strike the glass of the tube. The metal cross, however, stops them, and so there is seen at *d* a perfectly sharp shadow of the cross against the green glow of the wall of the tube.

Not all glasses show a green light : lead glasses give a blue, didymia glasses a red. A number of other bodies, minerals in particular, are, like glass, caused to fluoresce when they are struck by cathode rays.

The path of the cathode rays is quite independent of the position of the anode. The latter may be placed at the side, in front of, or behind the cathode, and the cathode rays are emitted in per- fectly straight lines from the cathode and perpendicular to it. If the cathode is made spherical like a con- cave mirror, the cathode rays, there- fore, meet at the centre of the sphere, which is called the *focus*, and after- wards diverge from one another.

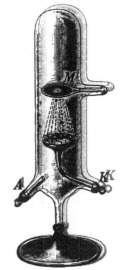

FIG. 6.

Metallic bodies do not fluoresce, but with them it may easily be shown that cathode rays heat bodies on which they impinge. In Fig. 6 a concave cathode is sealed into the tube at K, and its focus falls on a thin metal foil, M, which is placed at the correct distance in the tube. When the cathode rays are excited, the middle of the metal foil at once begins to glow. If the cathode rays from a concave cathode are allowed to fall on the glass of the tube so that the focus lies on the glass, the glass soon becomes soft and is then blown in by the pressure of the external air, so that the tube collapses.

Freely movable bodies inside a tube are set in motion by the cathode rays. In Fig. 7 a light wheel with mica vanes runs on two glass rails. As soon as cathode rays from the left or the right fall on the vanes the wheel turns to the right or the left and travels along the rails.

These properties of cathode rays led the English physicist Crookes to suppose that the molecules of gas in such a tube were flung off at high speed from the cathode, and that through the impact of these molecules on the walls heat was generated or else

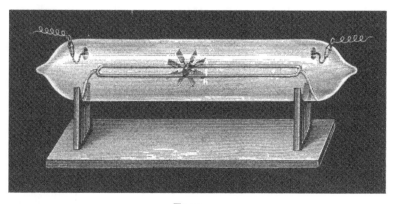

FIG. 7.

a convulsion of the particles took place, which showed itself in fluorescence. Crookes referred to a molecular bombardment as being the cause of the fluorescence and development of heat; the impact of these rapidly moving particles would also explain the motion of light bodies inside the tube.

But what sort of particles are they which are in rapid motion in the tube? That they cannot be, as Crookes assumed, the gas molecules themselves is demonstrated by another very surprising property of cathode rays. This is that the cathode rays are strongly deflected from their path by a magnet. To show this, a tube may be used of the form shown in Fig. 8, called a *Braun's tube*. In this the cathode is at K, the anode at A. The cathode rays proceed from K in straight lines and, in order that a quite narrow pencil of them may be obtained, they fall at S on a diaphragm of glass or metal, which has a small hole in the middle. They travel through this hole and fall on a screen, P, of calcium tungstate, a substance which is caused by them to fluoresce bright blue. Thus, in the middle of this screen there appears a small circular blue spot where the cathode rays impinge. If, now, one pole of a bar magnet,

FIG. 8.

say the north pole, is brought near the diaphragm, S, the blue spot is seen to be considerably displaced, which means that the cathode rays are deflected from their straight path. If the north pole is placed above the tube the blue spot moves backwards, if placed below it moves forwards ; if the pole is held on the front side of the tube the spot moves upwards, if behind the tube, it moves downwards. The cathode rays suffer the same deflection by the magnet as would an electric current which flowed to the cathode. Moreover, it is found that after deflection the spot is no longer round, but much extended in length, which shows that there are different parts in a cathode ray, some being more and some less strongly deflected by the magnet.

This experiment demonstrates that it cannot be the molecules of the gas that are travelling in the cathode rays ; for molecules of gas are neutral and would not be influenced by a magnet. If the whole idea of moving particles is correct, it must rather be electrically negatively charged particles which are moving in the cathode rays, for only such would

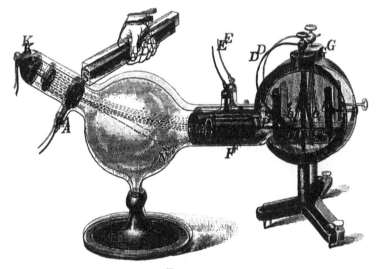

FIG. 9.

experience deflections in the direction which we found above.

As a matter of fact it may also be shown directly that the cathode rays carry negative charges with them. For this purpose we use the tube in Fig. 9. In this, K is the cathode and A the anode, which forms also a diaphragm with a small hole, so that the cathode rays fall directly on the spot, N. By means of the bar magnet, S, the rays are deflected so that they enter the side tube in which are two metal vessels, an outer one earthed through E, and an inner one connected with P. These metal vessels are called a *Faraday's cage*. By means of the wire, D, a gold leaf electroscope, G, is connected with P, and as soon as the cathode rays fall on the cage the leaves of the electroscope are seen to open. This shows that they are electrically charged, and on testing the charge (with a rubbed glass rod) it is shown to be negative. This demonstrates that it is electrically negatively charged particles which travel in the cathode rays.

The same explanation holds for the following important experiment which, as we shall see, allows

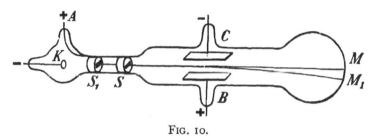

FIG. 10.

a quantitative measurement to be made. In Fig. 10 is shown a tube similar to Braun's tube, except that in it are placed two metal plates, B and C, forming together a condenser. Connections can be made from C to the negative and from B to the positive pole (or *vice versa*) of a battery with an E.M.F. of 100 or 200 volts. The cathode rays start from K, pass through the two diaphragms, S_1 and S, and produce a blue spot on a fluorescent screen at M, provided the condenser is not charged. If the latter is charged as indicated, the spot is displaced downwards to M_1, because the negatively charged parts of the cathode rays are attracted by the positive plate, B. The cathode rays suffer what is called an *electrostatic deflection*.

The particles which move in the cathode rays are therefore, without doubt, negatively charged, and it is reasonable to suppose that, as in electrolysis, we have here to do with ions, i.e. with electrically charged atoms. If that is the case the specific charge (see p. 32) of these particles must be less than, or at the most equal to, 96,494 coulombs/ gramme, and if we could determine the specific charge, we should be able to estimate the equivalent weight of the substance which carries the charge, and so could find out what the substance was.

The question is, then, What is the specific charge of the particles moving in the cathode rays, and a further question is, What is their velocity ?

Both these questions may be answered by carrying out the above experiments quantitatively, that is, by measuring the amount of the magnetic deflection and measuring the electrostatic deflection. As regards the electrostatic deflection, if a particle of unknown mass and of unknown charge travelling horizontally comes into an electrostatic field directed vertically downwards, its path is like that of a stone thrown horizontally and deflected by the earth, that is, it describes a parabola. From the amount of the deflection, the length of the condenser and the magnitude of the potential difference prevailing in it, the two unknown quantities, specific charge and velocity, are found in the combination, specific charge divided by the square of the velocity.

If the magnetic deflection is determined, we obtain from the curvature of the cathode ray and the strength of the magnetic field, the combination, specific charge divided simply by the velocity. It is easy to see that the two measurements afford a solution for the two unknown quantities in our problem, namely, 1, the velocity of the particles in the cathode ray, and 2, their specific charge.

First, as concerns the velocity of the cathode rays, this was always found to be enormous, some 100,000 kms. per second, roughly $\frac{1}{3}$ of the velocity of light. But this velocity appeared to be very different, according as the tubes were more or less thoroughly exhausted, and according as they were worked with a higher or lower voltage. In very highly exhausted tubes the velocity was higher, the magnetic deflection being smaller ; with a less perfect vacuum the velocity was lower and the magnetic deflection greater. The values found for the velocities in tubes which had been exhausted to different extents varied between $\frac{1}{5}$ and $\frac{1}{3}$ of the velocity of light.

In contrast with this, the value of the specific charge of the cathode rays in all kinds of tubes, whether more or less exhausted, always came out the same. And here there was a great surprise. For while one would have expected, from what was known of electrolysis, that the specific charge might be at the most equal to 96,494 coulombs/gramme, it was found that here it was almost 1900 times greater ; it came out to 177,000,000 ($1 \cdot 77 \times 10^8$) coulombs/gramme.

From this one might at first conclude that the electric charge is here combined with the atom of a substance whose equivalent weight is only 1/1900 of that of hydrogen, and that in these tubes the existence is surprisingly manifested of a totally new matter, which is 1900 times lighter than hydrogen, the lightest material hitherto known. But this strange conclusion does not stand closer examination. For it was found not to matter at all whether the rarefied gas in the tube was air, hydrogen, helium, oxygen, or any other gas ; the same value was always found for the specific charge. As it is impossible to assume that there is always, with all the different gases, the same admixture of a new unknown gas, this alone being the carrier of the cathode rays, we are forced to the conclusion that the cathode rays are electric charges which are not attached to any other mass at all, but that what we have here are free negative electrons, and that it is these electrons which constitute the cathode rays.

According to this, negative electrons would be tiny little bodies which have the elementary charge of electricity, but whose mass is only some 1/1900, more accurately 1/1835, of the mass of the hydrogen atom. But they would be material bodies like hydrogen.

However, further consideration and further experiments show that they are not material bodies like hydrogen. It may be demonstrated that they have no real mass at all but that their mass is only *apparent*.

In order to explain this somewhat difficult subject, we must examine what it is that leads us to suppose that a body has mass. We reach this conclusion on two grounds. Firstly, we know that all bodies are attracted by the earth and thus possess weight, and we feel this weight when we lift things up. The mass of all bodies on the earth is exposed to gravity, it is *heavy mass*. But suppose there were no gravity —(we might either imagine it suspended or we might think of a body placed on the line joining the centres of the moon and the earth 45,700 miles from the surface of the earth, where the attraction of the earth does not affect it because it is balanced by the attraction of the moon)—should we in that event still have experiences which would lead us to credit bodies with having mass ? Undoubtedly, and for the following reasons. To set in motion a body which is at rest a considerable force is needed. That is evidenced, for example, by a horse drawing a cart. Once the body is in motion with constant velocity, a much smaller force (none at all if there were no friction) is required to maintain this motion. So it requires force to move anything which is at rest (or, more generally, to accelerate anything). Inversely, if we have kept a body in motion by means of a force and then suddenly remove the force (e.g. when a driver on an electric tram suddenly cuts off the current), the body does not come to rest at once, but continues to move with the same velocity (and that indefinitely, assuming there were no friction to dissipate the velocity). These two properties of bodies depend on their having mass, and we call them inertia effects. The mass of a body is thus *inert mass*. It is from inertia that we recognize mass, inertia is for us the characteristic of mass.

But now we can see that an electron in motion shows the phenomena of inertia merely on account of its charge, even though it is not combined with a mass at all. For a charge in motion, such as a moving electron, is an electric current. But we know from the induction effects discovered by Faraday that, whenever a current begins to flow, i.e. when the electron begins to move, the magnetic forces which are propagated through the ambient ether produce an *extra-current* which opposes the original current. Force is therefore necessary to set the electron in motion. Inversely, when a circuit is suddenly opened, there results, for the same reason, an extra-current which tends to prolong the motion. So the electron continues to move, even when the force which kept it moving has stopped. These two properties of an electron caused by the extra-current (self-induction) correspond closely with the inertia of a mass. We must say that an electron has the property of inertia simply on account of its charge even though it has actually no mass. So we say that an electron has *apparent* mass. This semblance of mass is produced by the charge, through the magnetic effects of a charge in motion on the surrounding ether. It may be shown mathematically that the apparent mass of a charge is greater, the smaller the volume occupied by the charge is. For if we suppose the volume spherical, the magnetic lines of force which result are the nearer to the charge, and so the more effective, the smaller the radius of the sphere. The complete calculation shows that the apparent mass of an electron is equal to $\frac{2}{3}$ the square of its charge (in electrostatic units) divided by its radius and by the square of the velocity of light— provided the velocity of the electron does not approximate to the velocity of light.

From this we can calculate the size or radius of an electron. The mass of an electron is $1/1835$ of the mass of a hydrogen atom, and since the latter is equal, as on page 18, to $1·64 \times 10^{-24}$, the mass of an electron

$$= \frac{1·64}{1835} \times 10^{-24} = 9·0 \times 10^{-28} \text{ grms.}$$

The charge of an electron, according to page 33, is equal to $4·74 \times 10^{-10}$ electrostatic units. So the radius of an electron is

$$\frac{2}{3} \times \frac{22·5 \times 10^{-20}}{9·0 \times 10^{-28} \times 9 \times 10^{-20}} = 1·9 \times 10^{-13} \text{ cms.}$$

The *radius of an electron has the dimension* $1·9 \times 10^{-13}$ cms. We remember from page 15 that the radius of an atom is of the order of 10^{-8} cms. Let us illustrate the ratio by an example. The radius of the earth is about 6350 kms. If we suppose an atom of hydrogen magnified so that it occupies the whole volume of the earth, an electron will have a radius of only 120 metres, and so will correspond in size with no more than a large church. Just as this is minute compared with the whole earth, so is an electron minute compared with an atom.

The foregoing considerations do no more, at first, than render it possible that the mass of an electron may be *only* apparent. There might still be real mass or matter also combined with the electron. But such is not the case, as further investigations show. For it appears from the theory that the apparent mass of an electron has the above given constant value only if the velocity of the electron is small compared with the velocity of light, as is the case with cathode rays where it is not more than, say, $\frac{1}{5}$ or $\frac{1}{3}$ of the velocity of light. But if the velocity of an electron becomes much greater and approximates to that of light, then its apparent mass must also be greater, and if it attained to the velocity of light it would be infinitely great. Now, as will be explained in more detail in the next chapter, we know negative electrons, in the β-rays of radium, whose velocity is much greater than that in cathode rays, and it has, in fact, been shown experimentally that their mass is always greater the more closely their velocity approaches to that of light. It was found possible to ascertain this relation by determining their specific charge, i.e. the ratio of their charge to their mass. If the mass increases, the specific charge must become smaller and smaller the more nearly we approach the velocity of light. Experiments on the dependence of the specific charge on the velocity gave, for example, the following :—

Velocity.		Specific Charge.	
$1\cdot00 \times 10^{10}$ cm./sec.		$1\cdot77 \times 10^8$ coulombs/gramme.	
$1\cdot5 \times 10^{10}$	,,	$1\cdot77 \times 10^8$	,,
$2\cdot36 \times 10^{10}$	,,	$1\cdot31 \times 10^8$	,,
$2\cdot48 \times 10^{10}$	,,	$1\cdot17 \times 10^8$	,,
$2\cdot59 \times 10^{10}$	,,	$0\cdot97 \times 10^8$	,,
$2\cdot72 \times 10^{10}$	,,	$0\cdot77 \times 10^8$	,,
$2\cdot83 \times 10^{10}$	,,	$0\cdot63 \times 10^8$	,,

As the velocity increased up to 94 per cent. of that of light $(3 \times 10^{-10}$ cm./sec.) the specific charge decreased to about one-third, so that the apparent mass increased about three times. Accordingly, it may safely be concluded that *an electron has only apparent and no real mass.*

Such is the evidence which has been adduced for the existence of negative electrons, and since in electrolysis we obtain a positively and a negatively charged ion from a molecule, it is certain that a molecule must contain at any rate negative electrons besides the material atoms.

But now what about positive electricity or positive electrons? In the passage of a current through evacuated tubes under certain circumstances there occur not only the negative cathode rays, but also rays with a positive charge which are known as *positive rays.*[1] If, as in Fig. 11, there is placed inside the tube a cathode with holes or slits, i.e. canals, in it and in the lower part of the tube an anode, cathode rays travel downwards from K into the space where the anode is

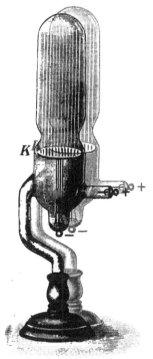

FIG. 11.

and excite green fluorescence in this lower part of the glass. But, in addition, other rays are seen to proceed upwards from the canals: they are of a reddish-brown colour in the case of air; these are the positive rays. That they carry positive charges

---

[1] In German writings the name " canal rays," given by their discoverer Goldstein, is almost exclusively used, but in English papers the term " positive rays " has become general, in view of its use by the Cavendish laboratory school: the word " Kanalstrahlen " is accordingly so translated.—TR.

with them is shown by the fact that they are deflected by a magnet and by an electrostatic field, but in precisely the opposite direction to the cathode rays.

By measuring the extent of the magnetic and electrostatic deflections, it was found possible, as with the cathode rays, to give an answer to the two important questions—" What is the velocity of the particles travelling in the positive rays, and secondly, What is their specific charge ? " The velocity was found to vary but to be always much smaller than for cathode rays. While in the latter case it was from $\frac{1}{5}$ to $\frac{1}{3}$ of the velocity of light, it is here about 1/500 to 1/1000 of the velocity of light, i.e. several hundred times less than for cathode rays. Of still more importance was the answer to the question as to the amount of the specific charge. Without exception this was found to be less than, or at the most equal to, 96,494 coulombs/gramme, whereas for cathode rays, as shown above, it was 1835 times greater. The first deduction from this is (see p. 33) that the electricity in the positive rays is not free but is invariably combined with ordinary atoms, in other words, the positive rays are made up of ions. Further, with cathode rays the value of the specific charge always came out the same whatever gas was in the tube : it is quite otherwise with positive rays. The figure for the specific charge is entirely different according as air, hydrogen, helium or oxygen is in the tube, which is evidence that the positive electricity is here always combined with atoms.

But a closer investigation of the positive rays has shown that in every case, whatever gas is in

the tube, there are several different kinds of carriers
of the positive electricity. The deflection experiments with positive rays are sometimes made not
by subjecting such a ray alternately to magnetic
and electrostatic deflection, but rather by allowing
the magnetic and electrical forces to operate on it
simultaneously in such manner that the deflections
due to the two forces are in directions at right angles
to one another. If the positive ray is allowed to
act on a photographic plate (inside the tube) a
deflected image is obtained making an angle with
each of these two directions ; by measurement of

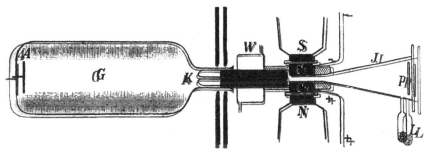

Fig. 12.

its position, both the magnetic and the electrical
deflections can be determined, and hence the specific
charge and the velocity of the particles of the positive
ray may at once be calculated. As the velocities of
the different particles of a positive ray are unequal,
a small circular bundle of rays gives, on deflection,
not a circle but a curve, actually a parabola. An
apparatus used by J. J. Thomson and Aston for
these experiments is shown in Fig. 12. A glass tube,
G, has an anode sealed in at A, and a cathode cemented
in at K. The latter consists of a metal cylinder
having in it a fine " canal." The forward part of K

is of aluminium with a rounded front surface, the rear part is of soft iron. The " canal " opens into a funnel-shaped glass tube, J, cemented on to it, which contains a photographic plate at P. Water running through the vessel, W, prevents the lute of the tube from becoming warm. The particles travelling through the " canal " are acted upon by an electro-magnet, NS, on the pole pieces of which are placed two insulated condenser plates, C_1 and C_2, which are in connection with the positive and negative poles of a battery. The electro-motive force

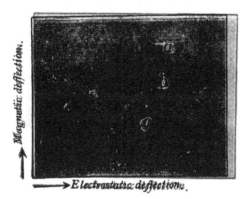

FIG. 13.

between C_2 and C_1 deflects the positive ray upwards (towards the negative plate) : the magnetic force between N and S deflects it forwards. There is thus obtained on the plate, P, a parabola extending from the centre upwards and forwards, if the positive ray has a fixed specific charge, i.e. if the positive charge is combined with one definite atom or molecule. If, however, as is actually the case, there are several different masses with a positive charge in the positive ray, several distinct parabolas are obtained on the photographic plate. Thus Fig. 13 shows an exposure in hydrogen where three distinct parabolas are revealed, Fig. 14 an exposure in a mixture of gases, with a large number of parabolas.

Whatever the gas in the tube is, there are always found different carriers of the positive electricity. In hydrogen it is found, for example, that there are positively charged atoms H^+ and also positively charged molecules H_2^+; the parabola, *a*, in Fig. 13 is due to H^+ atoms, the parabola, *b*, to H_2^+ ions. There seem to be even ternary compounds of hydrogen which are positively charged ions, that is, of the form H_3^+. Moreover, it has been shown that

FIG. 14.

among the positive rays there are also negatively charged rays which do not occur until the positive rays have been first produced. Thus there have been observed, e.g. negatively charged hydrogen atoms H^-. Similarly, there have been found positively and negatively charged oxygen atoms O^+ and O^-, also carbon atoms with a single positive or negative charge C^+ and C^- (parabola, *c*, in Fig. 13 comes from C^+ atoms), and some with a double positive charge C^{++}, and also positively or negatively charged molecules of carbon C_2^+ and C_2^-. By means of these positive ray photographs an insight is thus obtained into the different combinations into which atoms of electricity enter with atoms or molecules of material bodies. Positive ray photography serves thus for the chemical analysis of gases and is therefore also known as *positive ray analysis*.

This method of investigation has recently made extraordinary advances through the work of the English experimenter, F. W. Aston. In order that the images given by the different carriers of the positive charges should be really well separated

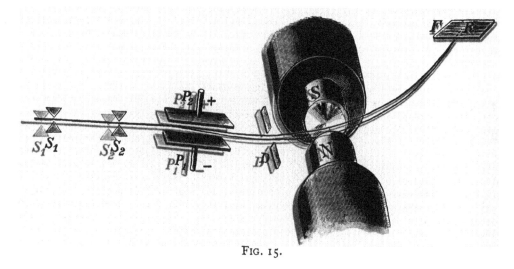

FIG. 15.

even though there is little difference between the masses of the carriers, the incident beam of positive rays must be extremely narrow, so that a very fine " canal " must be used. As a result, however, the intensity of the deflected rays becomes very small unless measures are taken to make rays with similar carriers but of different velocity come to the same spot. Aston succeeded in doing this by means of an apparatus which he calls a *mass spectrograph*, the principle of which is illustrated in Fig. 15. The positive rays pass through two very narrow parallel slits S₁ and S₂ and then arrive at the inclined condenser P₁ P₂, the lower plate of which is negatively charged. Imagine a number of positive rays having the same carriers, i.e. the same specific charge, but with different velocities. The more rapid rays will be less deflected in the condenser and the slower more deflected so that a beam is formed which is limited by a slit, D. These rays of different velocity now come into a magnetic field produced by two poles, a north pole, N, in front of the plane of the paper, and a south pole, S, behind it. In this magnetic field the rays are deflected upwards, that is, in a

FIG. 16

direction contrary to that in the electric field. The slower rays will be more sharply curved, the more rapid rays less curved. The former, therefore, describe a circle of smaller radius and the latter one of larger radius, and it may be arranged that these rays coincide at a point R.

Consequently all the positive rays with a certain carrier unite at a point R on a photographic plate, F, and all the rays with another carrier (with a different specific charge) at another point. If gases with known carriers are used for calibration, the specific charge, or its reciprocal the mass, of the carriers can be found corresponding with any point on the plate.

In Fig. 16 I is shown one of these mass spectrograms as obtained by Aston on putting carbon monoxide into the tube in which hydrogen and hydrocarbons were also present. On this spectrogram a series of five points is seen, corresponding with molecular weights of 12, 13, 14, 15 and 16, and due to the carriers C, CH, CH_2, CH_3 and CH_4. In II in the Figure, taken with another strength of magnetic field, these five points are seen closer together at the left, and further to the right are other points corresponding with equivalent weights of 24 to 28 ; the strong line 28 is due to CO, the others to hydrogen compounds. Calibration and measurements of such a plate thus give quite directly the mass of the carriers in the positive rays. This method has become of particular importance because by its means it has been possible to find out for certain whether a number of elements are really elements or are composite, a question to which we shall return in the third lecture.

For the present we may make two deductions from these investigations on positive rays. Firstly, free positive electrons have not been found, nor have they ever, we may add, even with other experimental arrangements. We might state the net result of all the experiments up to the present thus : while negative electricity may exist alone in the free state uncombined with matter, this is not the case with positive electricity. Positive electricity is always and universally associated with atoms of ordinary matter.

Secondly, we perceive that the combinations of atoms with one another to form molecules and with electricity to form ions are more varied than we first thought. There are positively and negatively charged atoms of the same kind, there are also some with more than one positive or negative charge : moreover, besides the neutral molecules there are also positively and negatively charged molecules with one or more charges.

A complete theory of the atom will have to take account of these manifold possibilities for the actual behaviour of atoms.

LECTURE III

THE DISINTEGRATION OF ATOMS IN RADIO-
ACTIVE SUBSTANCES—THE NUCLEUS THEORY
OF ATOMS

AT the beginning of the present century it
was discovered first by Becquerel and then
by P. Curie and Madame Curie that some
chemical elements have radio-active properties, i.e.
that they continually emit rays of a special kind.
Later investigation has led to the recognition of a
series of important properties of these rays and of the
elements from which they arise, properties which
are of particular importance for the theory of the
atom.

The first radio-active substance found by Becquerel was uranium, which has the property, both
in the metallic state and in all its compounds, of
blackening photographic plates even when they are
wrapped up in black paper : it also causes many
bodies to fluoresce and, finally, it ionizes air and
other gases or renders them conducting, so that
electrically charged bodies do not retain their charge
when uranium is brought near them, but are dis-
charged into the air which is then a conductor.
These effects are produced by rays emitted by the
uranium : uranium is what is called a *radio-active*
substance. Since, as we have remarked, all chemical

compounds containing uranium as well as the metal itself show these properties, radio-activity must be a property of the atom of uranium : it is an *atomic property*.

Much more marked effects, but of the same nature, are shown by radium, a substance which was separated, as is well known, by Mme. Curie in small quantities, after much trouble, from Joachimsthal pitchblende, which contains uranium. Radium, too, shows radio-activity both in the metallic state and in all its chemical compounds ; of the latter the chloride, bromide, and nitrate are principally prepared and used in practice. Here, again, it is the atoms that are radio-active. The atomic weight of radium was found by Mme. Curie to be 226. It is thus one of the heaviest atoms ; the atom of uranium is, however, heavier still, its atomic weight being 238·2.

Yet a third body of very high atomic weight, thorium, atomic weight 232·15, was found to be radio-active and finally another substance, actinium, was isolated in small quantities from pitchblende ; this, too, is radio-active, but its atomic weight has not yet been determined : it is in any event very high, estimated as about 227.

It is accordingly elements with the highest atomic weights which show the emission of radio-active rays. Detailed investigation of the radiation from all these substances has shown, however, that it is a mixture, three types of rays being present which are of different natures. These types may be separated with the aid of a magnet. For if there is in K (Fig. 17) a particle of a radium salt as a source of radiation, and a powerful magnet is placed with its north pole in front of the vessel, certain of these rays are deflected from their straight path and others are not. Of the rays which are deflected part are strongly deflected to the right ; these are called Beta-rays (β-rays). Another set of rays is weakly deflected to the left ; they are called Alpha-rays (α-rays). The third portion of the rays is not deflected : these are known as Gamma rays (γ-rays). With the rays which are not deflected, which behave

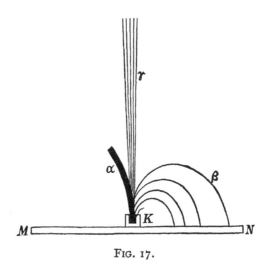

Fig. 17.

like X-rays, we shall deal in the next chapter ; here it may be mentioned that they are usually the most active. For they alone pass through fairly thick layers of solid, liquid, and gaseous substances, and when, as is usually the case, the radio-active material is enclosed in a vessel, it is consequently only these which penetrate the walls of the vessel and exhibit their effects on the outside. The medicinal application of radium depends almost exclusively on the γ-rays.

Since the α- and β-rays are deflected by a magnet they must consist, according to our previous explanation, of electrically charged particles. From the direction of the deflection it is seen that the α-rays are positively, and the β-rays negatively charged particles. Here again, as with the cathode rays and positive rays, the most important question is, what is the nature of these particles? This question and that as to the velocity of the particles was answered, just as for the other rays, by quantitative determination of the magnetic deflection on the one hand, and of the electrostatic deflection of the rays on the other hand. For from these two measurements both the velocity and the specific charge of the particles could be determined (see p. 42).

In the case of the β-rays the measurements showed that the specific charge was about 1.7×10^8 coulombs/gramme, of the same order, that is, as for cathode rays. Just as in the latter case, we must conclude, therefore, that it is actual negative electrons which form the β-rays. But the velocity of the radio-active β-rays was found to be different for different radio-active substances, and was sometimes immensely greater than that of cathode rays. In many cases the velocity approached to within a few per cent. of the velocity of light. We have already seen on page 48 that in this case the specific charge is less and that this is evidence for the mass of negative electrons being only apparent mass.

The nature of the α-rays with their positive charges is a question of peculiar interest at this stage. The α-rays carry nearly 90 per cent. of the total energy of radiation of a radio-active substance, so that most of the loss of energy occurs by their means. On the other hand, the α-rays are absorbed by very thin layers of foreign matter ; they do not penetrate even through thin paper or an aluminium foil 0·02 mm. thick. In air they attain distances of a few centimetres, 3-7 cms., according to their velocity, and then *suddenly* stop ; this distance is known as the *range* of the α-rays. It is an extremely striking, and at first inexplicable, fact that these α-rays should *suddenly* stop after traversing a path of a few centimetres. Up to this point of their route their capacity for ionizing gases and acting on photographic plates is quite unimpaired, and then they suddenly stop. They do not stop because they are gradually weakened by absorption, but there is just a sudden disappearance.

To fathom the nature of the α-particles it was again necessary to determine their specific charge. The same measurements which afforded a value for this gave also their velocity, which was found to be 50-100 times as high as that of the positive rays. Direct measurement of the specific charge gave the figure 48,200 coulombs/gramme which is of the same order as that of ions in electrolysis. The positive charges are here therefore combined with ordinary matter just as in electrolysis and in positive rays. As the specific charge of an atom in electrolysis is equal to 96,494 divided by the equivalent weight, it would follow that the unknown matter had an equivalent weight of about 2. But we do not know any atom with an equivalent weight of 2. The argument assumes, however, that the atom carries only a single elementary charge, an assumption which requires to be proved; it has, in fact, been shown to be incorrect. The distinguished English physicist, Rutherford, who has contributed most largely to our knowledge and understanding of radio-active phenomena, has demonstrated by an elegant method that each particle carries two positive elementary charges. Hence the equivalent weight of the particle in question must be four. Now we know a substance, helium, whose equivalent weight (and atomic weight) is four; whence the important conclusion is reached that *the α-particles are nothing but helium atoms with two positive elementary charges.* That radio-active substances emit doubly positively charged helium atoms in addition to negative electrons is a conclusion which has been confirmed by a series of experiments. In sealed vessels containing radium with, at first, no trace of helium, the presence of the latter is clearly demonstrated after some time by its very characteristic spectrum. The helium has been developed out of the radium.

Mme. Curie found, soon after the discovery of radium, that all bodies which were near radium became themselves radio-active, though only temporarily. A piece of paper, glass, wood or metal which is near radium, itself emits rays which render the air conducting. But these bodies retain the property only for a relatively short time, after which they again become inactive. Their activity is known as *induced activity*. The secret of it was brought to light by a discovery which Rutherford made, in the first place, with thorium, but which was also found later to hold for radium. He found out that a gaseous substance, called by him *emanation*, is produced from them, which is itself radio-active, causes bodies to fluoresce and ionizes air. This emanation does not retain its activity unaltered, but gradually loses it. The activity of radium emanation decays in 3·85 days to the half of its original value, as was proved by measurement of its ionizing power; thorium emanation falls to one half in 54 seconds and actinium emanation in 3·9 seconds. These times, known as *constants of decay* or *half-value periods*, are quite characteristic and are always found to be the same under all circumstances for the emanation of a radio-active body. They do not depend on whether the measurement is made immediately after evolution of the emanation or later, or on whether much or little material is used. The emanation is accordingly a radio-active material too, but one which, unlike the first-mentioned, radium, thorium, etc., which keep their activity for ever, gradually loses its activity : it is a substance with temporary activity. Induced activity also exhibits decay; but in contrast to emanations, the values of the decay constant are found to be very different according as the induced activity has lasted for a shorter or longer time.

The striking fact that there are temporarily radio-active bodies led to the theory, first proposed by Rutherford, that the atoms of radio-active bodies are not unalterable but disintegrate, the *theory of the disintegration of atoms*. According to this, forces are actually at work in radio-active substances, and that without our intervention, which loosen the coherence of the atom and burst the atom to pieces.

The unalterability and indivisibility of atoms are here, in fact, refuted.

By the loss of an α-particle from an atom of radium there is produced an atom of a new substance, that of the emanation. But this atom again disintegrates, with the loss of an α-particle, and now forms an atom of the substance which, collectively, is called induced activity. The reason for the indefiniteness of the decay constant of the induced activity is that it is not a simple substance : a series of different substances is produced from it, which therefore exist to some extent concurrently, and will be in different relative amounts according to the time which has

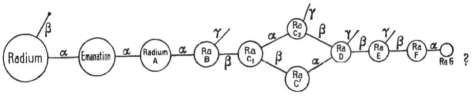

FIG. 18.

elapsed since it was first formed. This series of substances formed from the emanation is designated radium-A, radium-B, etc., to radium-F. An atom of one of these bodies is produced from the preceding one by the emission of α- or β-particles or of both, sometimes with the accompaniment of γ-rays (X-rays).

The gradual formation of these separate substances is illustrated by Fig. 18 for the case of radium, according to recent experiments ; the nature of the ray which produces any atom from the preceding one is also shown. The substances radium-A, B, and C_1 have very small half-value periods, namely, 3, 26·8, and 19·6 minutes, so that they are substances

5

which rapidly disappear and cannot be recognized after a few hours. From some of the atoms of radium-C_1, there are emitted first α-particles yielding radium C_2, which by loss of a β-particle becomes radium-D ; from other atoms of radium-C_1, however, a β-particle is first shot out leading to the formation of radium-C', which then loses an α-particle and again forms radium-D. At radium-C_1 there is what is called *branching*. Radium-D has, however, a very large half-value period, 16·5 years, and is thus a very stable body which is only gradually transformed to radium-E and F with emission of barely recognizable

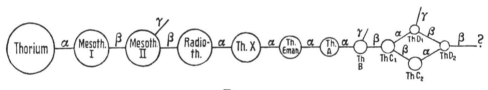

FIG. 19.

β-rays. These two are also fairly long-lived, for they have half-value periods of 5 and 136 days. Radium-F is absolutely identical with polonium, which was separated by Mme. Curie from pitchblende at the same time as radium, but which, as we now see, is only a descendant of radium. From radium-F α-rays are emitted resulting in a substance radium-G with the nature of which we shall deal later.

In a similar manner the development of atoms from thorium has been established as represented in Fig. 19. Here again there is branching at thorium-C. The times of decay of the different members of this series are as follows :—

Substance.	Time of Decay.
Thorium	——
Mesothorium I	6·7 years.
Mesothorium II	6·2 hours.
Radiothorium	1·905 years.
Thorium-X	3·64 days.
Thorium emanation	54·53 hours.
Thorium-A	0·14 seconds.
Thorium-B	10·6 hours.
Thorium-C_1	60·8 minutes.
Thorium-D_1	3·1 minutes.

The actinium series is of less importance. On the other hand it is rather interesting that radium is a direct descendant of uranium. For it has been found that uranium is transformed through an intermediate body, ionium, which has an extraordinarily long half-value time (about a million years), into radium which itself has a half-value time of some 2000 years.

All these substances, uranium, radium, thorium, actinium and their products emit α- and β-rays, and in every case the β-rays are the same and the α-rays the same. All β-rays, from the most different substances, are negative electrons which are only distinguished from one another by the difference in their velocities. And all α-rays, from whatever substance, are helium atoms with a double positive charge, which are only distinguished by the fact that they have different velocities and cover different distances in the air before they cease to ionize it, i.e. they differ only in their range.

But hence it follows of necessity that the atoms

of these substances, uranium, radium, thorium, actinium and their derivatives must contain helium atoms and negative electrons as constituents. In these very heavy atoms the ultimate constituents are therefore known from these investigations. It does not necessarily follow from what has been said that a radium atom, for example, must consist *only* of helium atoms and negative electrons; it might have other substances also as constituents, e.g. hydrogen atoms. But hitherto no evidence has been obtained either for or against this possibility.

The expulsion of α-particles from a piece of radium or other radio-active substance may be rendered *visible* by a most interesting experimental arrangement. The different types of rays, both α- and β- (and also γ-), render the air through which they travel electrically conducting, they ionize the air as we say. For if particles of air or other gas are struck by such rays, the gas particles are broken up and there are formed from a molecule of gas positive and negative ions, gaseous ions. As to the nature of these ions we are not at present further interested. The fact that gases become conductors, a change which can be followed by measurement, has provided for a long time the chief method for the investigation of radio-active phenomena.

Now these gaseous ions have a peculiar property; they can act as what are called *condensation nuclei*. If some water is placed at the bottom of a closed vessel containing air, water vapour is formed, which mixes with the air and is just as invisible as the air. According to the temperature of the vessel there is a larger or smaller amount of water in the form of vapour ; the water vapour is said to be saturated, there being just so much vapour present as corresponds with the temperature and volume of the vessel. If now the temperature of the vessel is suddenly reduced, there corresponds with this lower temperature a smaller amount of saturated water vapour ; consequently the excess of vapour must be precipitated in the form of a cloud of droplets of water. This should occur even with a very small reduction of temperature ; but, as a matter of fact, on account of surface tension, such is not the case, a fairly considerable reduction of temperature being necessary to cause the formation of a mist from the vapour in pure air. The formation of cloud is, however, considerably facilitated, as was found out some time ago, if dust is present in the air, which is usually the case if the air has not been carefully filtered. The particles of dust act as condensation nuclei and drops of water are readily formed on them by even a slight reduction of temperature. Now it has been found that gaseous ions act just like dust ; water vapour is precipitated on gaseous ions as liquid water by even a slight reduction of temperature. In order to bring about a slight lowering of temperature rapidly in air, the vessel in which the air is contained in these experiments is suddenly put into communication, by opening a tap, with another vessel at a lower air pressure. The sudden efflux and increase of volume causes cooling of the air, and a cloud is at once formed if dust nuclei or gaseous ions are present.

This property of gaseous ions was utilized by the English physicist, Wilson, in a very beautiful experiment to render the paths traversed by α- or β-particles directly visible, capable in fact of being photographed. We are here interested only in the paths of the α-particles. He allowed α-particles from a piece of radium to enter the vessel which contained the water vapour, which he called the

FIG. 20.

cloud-chamber. These particles ionized the air in the cloud-chamber and formed along their route a large number of gaseous ions. On suddenly cooling the gas in the manner described above, water vapour is precipitated on these ions and thus a streak of mist appears in the cloud-chamber showing directly the path of the α-particles. By suddenly illuminating the streak of mist at the correct moment this may even be photographed. Figs. 20 and 21 reproduce some photographs taken by Wilson of the α-rays emitted by a piece of radium. They correspond closely with the picture which previous experiments on the α-rays had led us to draw. In particular

FIG. 21.   FIG. 22.

they show that each ray has only a certain range and that it then *suddenly* stops. On closer examination, however, some of these α-rays show a strange peculiarity which is of eminent importance for our further discussion. Thus it is seen in Fig. 21 that some of these rays exhibit, shortly before their termination;

a sudden sharp bend; the straight ray has at one point a small appendage at an obtuse angle. Not all rays show this, only some of them. In Fig. 22 two of the rays in Fig. 21 are shown magnified, one of them exhibiting the bend, the other not doing so. The α-particles collide with particles of air during their flight and ionize them, but in general they persist in their straight path : their energy is so great that collision with particles of air does not deflect them from their path. But now and then some of them suddenly experience a marked deviation and soon afterwards cease to exist. What may be the reason for this remarkable phenomenon ?

A further observation made in connection with α-particles is related to that above described. When an α-particle, i.e. a helium atom with a double positive charge, passes through a thin metal foil, it experiences as a result a deflection from its path, usually small ; consequently a narrow pencil of α-rays exhibits some broadening or *scattering* after passage through a thin metal foil. The reason for this is easily seen. For since the atoms of the metal foil themselves contain positive and negative charges, the positively charged helium atom will be somewhat deflected by them, first in one direction, then in another, but on the whole only slightly. It has been calculated from observations on scattering that α-particles experience, on the average, a deflection of only 1/200 degree by an atom of gold. Depending on the manner in which an α-particle meets an atom, the deflection is naturally sometimes greater, sometimes less, according to the laws of probability, but on the average it is only of the stated small order. It was noted, however, in these observations that now and then deflections occurred which were extraordinarily great, 90° and more ; on the laws of probability these should never have been observed, for they would have happened only once in millions or billions of cases, whereas actually they took place once in some 8000 times. These large deflections gave the impression that the α-particle was, once in a while, directly reflected from an atom of the metal foil. Similar scattering of α-particles occurred, not only in metal foil but also in gases, and in certain cases was of just as improbable a magnitude. This is, it will be seen, precisely the same phenomenon as that shown in Wilson's photographs Figs. 21 and 22 ; here, too, an α-particle appears once in a while to be reflected directly from an atom, and shows in consequence a crooked path.

After deliberating on the cause of this remarkable phenomenon, Rutherford came to the conclusion that it could only be found in the constitution of the atoms of the layer in which the scattering occurred. For the magnitude of these high deflections is directly related to the atomic weight of the scattering substance. The greater the atomic weight (squared) of the scattering substance is, the higher are these extraordinary deflections. Hence the deflections, both the ordinary small ones and the rare large ones, could only arise from an α-particle shooting through an atom. In that case there must be empty spaces in every atom, and the first conclusion is that the space occupied by an atom cannot be completely filled with mass.

The curious phenomena of α-particles, their

sudden stoppage after traversing their range, their usually weak scattering and their unusually strong deflections are explained, as Rutherford showed, if we assume that an atom consists of a positively charged nucleus of extremely small volume around which, at a distance corresponding roughly with the atomic radius, are negative electrons sufficient to make the whole atom neutral. These negative electrons are either at rest or revolve round the nucleus like planets round the sun. An α-particle, whose dimensions must be supposed very small, will generally shoot through the atom somewhere in the interspace between the positive nucleus and the electrons, and will only experience a small deflection which is the resultant of the repulsion from the nucleus and the attraction by the electrons ; this is small because, in its flight through the atom, the α-particle moves at a relatively great distance from the nucleus and from the electrons. It may happen, however, that an α-particle flies directly towards a negative electron or, in rarer cases, directly towards the positive nucleus. In the former case it will combine with the negative electron and change from a helium atom with two positive charges to one with a single charge, and if it then meets yet another electron it will become a neutral helium atom. According to the velocity of the α-particle this will happen sometimes earlier, sometimes later. But once this neutralization of its charge has occurred the helium atom suddenly loses its property of ionization and thus is explained the sudden stoppage of the α-rays after they have traversed their range.

On account of the assumed smallness of the

nucleus it will very much more rarely happen that an α-particle is projected into the immediate neighbourhood of the nucleus of the atom. When this occurs there will be a strong repulsive force between the positive charges of the α-particle and of the nucleus, and the α-particle must be deflected at a sharp angle (more accurately in a hyperbola) from its path. In this manner are explained the two phenomena which are so striking and at first so mysterious, the sudden stoppage of the α-rays and their occasional sudden deviation, phenomena which at first appeared to baffle explanation.

Mathematical analysis of the process by which the α-particle is deflected by the nuclear charge shows that the deflection must depend on the magnitude of the positive charge of the nucleus, on its square in fact. When the experimental results were tested in the light of this theory, complete agreement was found and, from the observed degree of deflection, it was possible to calculate what the charge on the positive nucleus of the different atoms was, taking that of hydrogen as unity. It was then seen that the charge on the central nucleus of the different atoms was roughly equal in number to half the atomic weight, taking that of hydrogen as unity. Thus helium with an atomic weight of 4 showed a positive charge of 2 on the nucleus, carbon with an atomic weight of 12 one of 6, oxygen with an atomic weight of 16 one of 8, sulphur with an atomic weight of 32 one of 16.

This is a very surprising result and it has a still more surprising explanation. If all the elements are arranged according to the periodic system, i.e. in the order of their atomic weights, the place occupied by the above-mentioned elements is just half its atomic weight. For example, the first elements arrange themselves thus : 1 hydrogen, 2 helium, 3 lithium, 4 beryllium, 5 boron, 6 carbon, 7 nitrogen, 8 oxygen, 9 fluorine, 10 neon, 11 sodium, 12 magnesium, 13 aluminium, 14 silicon, 15 phosphorus, 16 sulphur, etc.

Thus the *ordinal number* in the periodic system of the elements [1] comes out equal to the positive charge on the nucleus of the atom or, in other words, the number of positive elementary charges which the nucleus possesses determines its ordinal number in the periodic system, and therefore determines also its atomic weight ; for the serial order of the elements in the periodic system is, in general (with only a few exceptions) the serial order of their atomic weights.

But this leads to a very new and surprising idea. The positive charge of the nucleus determines at the same time the mass of the atom. But this can only mean that the mass of the atom is a property which depends solely on its charge, i.e. *the mass of the atom is apparent mass*, just as the mass of the negative electron is apparent mass.

This is a deduction and a conclusion of extraordinary audacity ! It states nothing less than that the mass of bodies, which we have been accustomed, from our youth up, to regard as the most solid fact, is just an illusion. The first experiences which a child has, when he hits himself on the edge of a

---

[1] Or the " Atomic Number," as it is usually called.—TR.

table, give him the definite and indelible impression of the reality of mass. (Translated from the childish to the scientific, this experience states that a force is necessary to set a mass in motion, i.e. that mass has inertia). Mass appears to the child, and to us, as the first and plainest fact that we know of nature. And now we are explaining this first and plainest fact as a semblance. Not that we deny the effects of mass, as they are experienced by the child on the edge of the table or by the soldier who is hit by a piece of shrapnel. But we explain that this effect arises, not from a special something which we designate mass, but simply from the charges that the nucleus of the atom carries ; that this mass is hence nothing but a result of the charge and that therefore it is not the primary fact, self-evident, self-subsisting, but that it is secondary, a consequence of the charge, not subsisting without this charge. In a word, we say that mass is not something primary, in-dwelling in bodies, but that rather the electric charges are to be explained as the primary, and mass but a consequence of these charges. Anyone who has studied mechanics, scientifically or technically, will have had mass represented to him at the beginning of his course as something got from experience, which neither can be nor need be more closely defined but is given. That which was simply given we now deny and refer it to something else, to the electric charge ; no one could have suspected, when mechanics was in its fullest flower and perfectly developed, that charge should be the ultimate cause of mass.

But if the mass of an atom is merely apparent mass, we may at once obtain an idea of the dimensions of the positively charged nucleus. For according to the formula given on page 46, mass equals

$$\frac{2}{3} \frac{\text{square of charge (electrostatic units)}}{\text{radius of electron} \times \text{square of velocity of light}}$$

The mass of a hydrogen atom (see p. 18) is $1 \cdot 64 \times 10^{-24}$ grms. The elementary charge (see p. 33) is $4 \cdot 74 \times 10^{-10}$ electrostatic units.

Consequently, radius of the hydrogen nucleus

$$= \frac{2}{3} \frac{22 \cdot 5 \times 10^{-20}}{1 \cdot 64 \times 10^{-24} \times 9 \times 10^{-20}} = 1 \cdot 0 \times 10^{-16} \text{ cms.}$$

As we found (p. 47) that the radius of a negative electron is $1 \cdot 9 \times 10^{-13}$ cms., we see that the positive nucleus, which contains all the mass of the hydrogen atom, has a radius some 2000 times smaller even than an electron.

If we suppose again that the atom (whose order of magnitude is 10^{-8} cms.) is magnified so that it occupies the volume of the earth, of radius 6350 kms., then the nucleus of the hydrogen atom will have a radius of only 6 cms. corresponding thus with about the size of a child's ball, while a negative electron, at the same magnification, corresponds with the volume of a large church, its radius being 120 m.

If we could assume that all atoms, even the heaviest, consist only of a single nucleus, we should be able to calculate in the same way the radii of all the nuclei. For the number of elementary charges in the atom is fixed by its ordinal number in the periodic system, and its mass is given by the atomic weight. But as we know already that, at any rate in the heaviest atoms, those of radium, thorium, actinium, and uranium, there are helium nuclei present, we must regard these, and presumably most other atoms, as a conglomerate of a larger or smaller number of such nuclei ; we have only hitherto been able to recognize helium nuclei, but it is possible (and as we shall see is actually the case in many atoms) that they may also contain other nuclei, e.g. nuclei of hydrogen with a single charge.

Now if we must regard the central nucleus of the heavier atoms as a conglomerate, a juxtaposition of simple nuclei, this has a further consequence. A number of positively charged nuclei, be they nuclei of helium or of hydrogen or what not, cannot form a conglomerate of their own accord, for the different positive nuclei must mutually repel one another. In order that they may remain in juxtaposition it is necessary that the complete nucleus should contain, in addition, negative electrons to provide attraction and link up the different sub-nuclei. Thus in the heavier atoms, certainly in those of radio-active substances and probably also in many others, we must suppose the central nucleus to be made up of simpler positive nuclei combined with a larger or smaller number of negative electrons. Since, however, each negative electron in the nucleus diminishes the total positive charge by one unit there must be so many more positive sub-nuclei combined in order to give the atomic number of the element. For example, an element with an atomic number of 92 (uranium) would, if it were composed only of helium nuclei, contain 46 of them. For these to stay together there must also be present a number, and that an even number, of negative electrons.

With two negative electrons 47 helium nuclei would be required, with four negative electrons 48 helium nuclei and so on. From the fact that the atomic weight of uranium is 238·2 and that of helium 4, it would follow that a maximum of 59 helium nuclei and 26 negative electrons could be contained in the nucleus of the uranium atom. In order to be able to make definite statements about the constitution of the atom there are just as many possibilities to be tested by comparison with facts as there are in the chemistry of complicated molecules containing many atoms. We see before us a totally new field, the structural chemistry of the nucleus, analogous to the structural chemistry of the molecule, as a problem for the future.

According to this new conception, that which characterizes an atom and distinguishes one atom from every other is the nucleus, i.e. the positive charge which the nucleus possesses. A hydrogen atom is characterized by a nucleus with one elementary charge, a helium atom by one with two elementary charges. Around these nuclei revolve negative electrons in circles or ellipses, but they form no part of the nucleus. If there is no such electron within the range of attraction of the hydrogen nucleus we speak, in our modern phraseology, of a positive hydrogen ion H^+; this is then simply the nucleus. If there is one negative electron within its sphere of attraction we speak of a (neutral) hydrogen atom H. If there is a second negative electron in the sphere of attraction we speak of a negative hydrogen ion H^-. Similarly, with helium we may distinguish helium atoms with a double positive charge He^{++}, which are nothing but the nuclei or the α-particles: by the addition of successive negative electrons there will be formed from them helium atoms with a single positive charge He^+, then neutral helium atoms He, and finally the negatively charged helium atoms He^- and He^{--}. But all these different substances with different optical and electrical behaviours are helium, because they are all characterized by the same nucleus.

A distinction is necessary, in the heavier composite atoms, between the negative electrons which are situated in the nucleus itself, holding it together, which we will designate *nuclear electrons*, and the negative electrons which revolve about the nucleus at a smaller or greater distance away from it. Increase or decrease in the number of the latter does not alter the atom, the substance. The substance and its chemical behaviour are defined by the positive charge of the nucleus. On the other hand, an increase or decrease in the number of the nuclear negative electrons alters the atom, the substance, and changes it into a new substance with a different chemical behaviour, just as an increase of α-particles in the nucleus changes the substance. If we now follow again the radio-active transformations from this point of view, we shall find that we must extend our ideas in the case of the heavy atoms which contain nuclear electrons in the positive nucleus; for we shall see that two atoms, which we must consider to be chemically identical according to our conceptions, and which also have exactly similar chemical behaviour, may still have different atomic weights. This is in flat contradiction to what has hitherto

6

been supposed about the atom, for up till now it was precisely the atomic weight that was the chief characteristic of the atom, by which one kind of atom was distinguished from another kind.

For if a substance with a certain nuclear charge emits an α-particle, its nuclear charge becomes less by two units. But if it then emits a β-particle, i.e. a nuclear electron, its positive nuclear charge is reduced by one negative unit, i.e. is increased by one. And if it then emits yet another β-particle, its nuclear charge is again exactly the same as it was at first, so that the same atom, i.e. a chemically equivalent atom, is reproduced and yet there is a difference in the atomic weights of these two substances, the original one and that resulting from the loss of one α-particle and two β-particles. For the mass of the atom resides essentially in the positive nuclear charges, in this case in the α-particles ; the β-particles have only $1/1835$ of the mass of a hydrogen atom, and, therefore, contribute practically nothing to the total mass of the atom. Evidently the substance formed after emission of one α-particle and two β-particles has an atomic weight less by four units than the original. But, as the positive charge on the nucleus is the same in both, there is no chemical difference between the two bodies ; in particular they belong to the same place in the periodic system. The two atoms are not identical, in fact, we know the difference between them, viz. that they contain different numbers of α- and β-particles, but they are chemically inseparable.

Such elements, which have the same nuclear charge but yet differ in the number of positive *and* negative charges in the nucleus, are called *isotopic elements*. They occupy the same place in the periodic system and cannot be separated from one another by chemical means although they have different atomic weights. Such cases have been frequently observed in the investigation of radio-active substances and were first intelligible on the nuclear theory of the atom.

To work through an example, let us start from ordinary uranium (uranium-I) which has an atomic weight of 238·2 and whose nuclear charge, as we shall see in the next chapter, has been determined as 92. We observe that on emission of an α-particle the atomic weight diminishes by four units, and the nuclear charge by two units. On the other hand, the emission of a β-particle leaves the atomic weight (practically) unchanged but increases the nuclear charge by one positive unit.

The above-mentioned series of transformations from uranium through ionium into radium gives, accordingly, the following series; those atomic weights which have only been calculated are stated without decimals.

Transformation	Uranium-I $\xrightarrow{\alpha}$	Uranium-X$_1$ $\xrightarrow{\beta}$	Uranium-X$_2$ $\xrightarrow{\beta}$	Uranium-II $\xrightarrow{\alpha}$	Ionium $\xrightarrow{\alpha}$
Nuclear charge	92	90	91	92	90
Atomic weight	238·2	234	234	234	230

Transformation	Radium $\xrightarrow{\alpha}$	Radium emanation $\xrightarrow{\alpha}$	Radium-A $\xrightarrow{\alpha}$	Radium-B $\xrightarrow{\beta}$
Nuclear charge	88	86	84	82
Atomic weight	226·0	222	218	214

Transformation	Radium-C$_1$ $\xrightarrow{\alpha\beta}$	Radium-D $\xrightarrow{\beta}$	Radium-E $\xrightarrow{\beta}$	Radium F $\xrightarrow{\sigma}$	(Radium-G) $\rightarrow$
Nuclear charge	83	82	83	84	82
Atomic weight	214	210	210	210	206·0

In this series the nuclear number 84 is seen to occur with the elements radium-A and radium-F (polonium) which are thus isotopic elements. (Radium-C′, which is formed from radium C_1 by loss of a β-particle, is also isotopic with radium-A.) Again, the nuclear number 83 occurs with the elements radium-C_1 and radium-E, which are thus also isotopic. Finally, the elements radium-B, radium-D, and the unknown end-product of the series, radium-G, have a nuclear number 82.

As to the nature of this end-product, the conjecture has long been hazarded that it is lead, whose nuclear charge is 82 and atomic weight 207·2. But since radium-B and radium-D are isotopic with radium-G they must also be lead, or at any rate indistinguishable chemically from lead except for their atomic weights. It follows that there must be lead with different atomic weights according to its method of formation, e.g. that which consists of radium-D and that which consists of radium-G, of which the atomic weights are 210 and 206, whereas ordinary lead has an atomic weight of 207·2. This conclusion has been experimentally verified by accurate determinations due to Hönigschmidt, in which he actually found an atomic weight of 206 (lead from uranium) for lead contained in pure pitchblende, i.e. in a substance containing radioactive matter, instead of the usual value 207·2. The large differences which might theoretically occur in the atomic weight of lead—between 214 (radium-B) and 206 (radium-G)—can naturally not occur in practice since the radium-B, D, and G can only be present in very small quantities. If we include also the other radio-active series, we find that not only radium-B, D, and G, but also thorium-B and actinium-B are isotopic with lead.

It is not at present possible to give a complete exposition of these ideas, as in many cases the observations do not yet provide sufficient certainty as to the successive radio-active transformations.

On the other hand, we can now see that it may be possible for isotopes to occur not only among the radio-active but also among the ordinary elements. Two elements may have the same nuclear charge and absolutely identical chemical behaviour and yet have different atomic weights because the nucleus, though equally charged, has a different structure. We cannot separate two such substances chemically, but the atomic weight of the material which we regard as simple will, if it consists of two isotopes in different proportions, have a value intermediate between the atomic weights of the two isotopes. Here is a possible reason for the marked deviation of atomic weights from whole numbers (p. 6) which in many cases, e.g. chlorine 35·46, is quite certain, in spite of the fact that many other atomic weights are very nearly whole numbers. Physics provides, however, one means for the recognition of such isotopes, at any rate in the gaseous state. For if positive rays are generated in such a gas, the specific charge of the positive rays will be different according as the charge resides on the atom of one or the other isotope. It should be possible to recognize this by positive-ray analysis (p. 54) since the positive rays of the two isotopes will experience different deflections. Actually the method of positive-ray analysis has shown itself well suited to answer this question. The English physicist Aston has succeeded in demonstrating, by means of the mass spectrograph (p. 54), that in many cases apparently simple substances are composed of two or more isotopes. Neon, for example, which has an atomic weight of 20·2, was shown to be made up of two constituents with atomic weights of 20 and 22. In Fig. 23 is reproduced one of Aston's positive-ray photographs taken when 20 per cent. of neon was mixed with the carbon monoxide in the tube mentioned on page 56. In addition to the five lines before mentioned, corresponding with atomic weights 12–16, two more lines are clearly

FIG. 23.

shown with atomic weights of 20 and 22 ; these represent the isotopes of neon, the former being noticeably stronger than the latter so that there is only a small admixture of so-called metaneon (atomic weight 22) with the neon. Again, to the left of the five lines two others are seen, corresponding with doubly charged neon and metaneon, the specific charges of which are thus 10 and 11. In the same manner Aston showed that chlorine consists of two isotopes with atomic weights of 35 and 37, the former of which gives an image three or four times as strong as the latter. Fluorine, phosphorus, and arsenic yielded no isotopes but corresponded with atomic weights of 19, 31, and 75 ; with sulphur, atomic weight 32, it was doubtful whether it had an isotope, but boron showed two isotopes of atomic weights 28 and 29, and possibly also 30, bromine two isotopes with atomic weights 79 and 81 ; iodine (127) has no isotopes. With argon there was found, in addition to the substance with atomic weight 40, a small admixture of an isotope with an atomic weight 36. With krypton six isotopes were found with atomic weights 84, 86, 82, 83, 80, and 78, in the order of their intensity, with xenon five isotopes with atomic weights 129, 131, 132, 134, and 135. Mercury shows, besides an unresolved band of atomic weight 197–200, an isotope of atomic weight 202 and a weaker one 204. It is very interesting to note that Broñsted and Hevesy have succeeded in partially separating the isotopes of mercury by a special type of distillation, and that they found that mercury is composed of equal volumes of the pure elements 202·0 and 199·2. Among the alkali metals again, Aston showed that lithium is probably composed of two isotopes with atomic weights 6 and 7, while sodium (23) has no isotopes. Potassium consists of an isotope 39 with a small proportion of another with atomic weight 41 ; rubidium similarly has two isotopes 85 and 87, while hitherto no isotope of cæsium (133) has been recognized. Helium, carbon, nitrogen, and oxygen showed no isotopes. It is noteworthy that all these newly discovered elements have atomic weights which are exactly *whole numbers* if that of oxygen is taken as 16. Hydrogen, nevertheless, shows on this scale an atomic weight not of 1 but of 1·008.

If these new results are arranged according to the periodic system of the elements we obtain the following synopsis [1] for the elements examined, by means of which the respective data in the table facing page 7 may be corrected.

0.	I.	I.	III.	IV.	V.	VI.	VII.	VIII.
He 4·0	Li 6, 7	Be 9	B 10, 11	C 12	N 14	O 16	F 19	
Ne 20, 22	Na 23	Mg 24, 25, 26		Si 28, 29, 30	P 31	S 32	Cl 35, 37	
A 36, 40	K 39, 41	Ca 40, 44						Ni 58, 60
		Zn 64, 66, 68, 70			As 75	Se 74, 76, 77, 78, 80, 82,	Br 79, 81	
Kr 78, 80, 82, 83, 84, 86	Rb 85, 87							
				Sn 116, 117, 118, 119, 120, 122, 124	Sb 121, 123		I 127	
Xe 124, 126, 128, 129, 130, 131, 132, 134, 135	Cs 133							
		Hg 199, 202, 204		Pb 206, 210, 214				

In this unexpected manner the difficult riddle of the marked deviation of some atomic weights from whole numbers has been solved, and from this standpoint there is no longer anything essentially inconsistent with Prout's hypothesis (p. 6).

[1] A number of additional results obtained by Aston and by Dempster since the date of the German edition have been inserted in this list.—Tr.

LECTURE IV

X-RAY SPECTRA AND THE NUCLEUS THEORY OF THE ATOM

THE nucleus theory of the atom has received an unexpected but striking confirmation from recent investigations on Rontgen or X-rays.

Scientific research into X-rays made very little progress from the time of their discovery (1895) until the year 1912. It has been known since they were first discovered that there are X-rays of different kinds, which were roughly distinguished as hard and soft X-rays. The difference is that hard rays are little absorbed by a body such as a piece of aluminium of a certain thickness, whereas soft rays are absorbed by the same body to a marked extent. Every X-ray bulb emits both hard and soft rays in different proportions, it gives a *non-homogeneous* radiation, not a radiation which is of only one quality such as is called homogeneous. As to the nature of these rays, the question which is of most importance to physicists is whether they are corpuscular rays or a process similar to light rays ; no direct answer could be obtained to this question though the balance of probability pointed strongly to the second of the two possibilities. But though X-rays do in fact appear to have many properties in common with light rays, yet there were on the other hand such important differences between them that no definite answer could be given.

Light consists, as is well known, of a wave motion. It is propagated *in vacuo* with a velocity of 300,000 kms. per second. The wave-length of rays of light is extremely small, and is different for different colours. Red rays have a wave-length of about 760 millionths of a millimetre, the violet about 380 millionths of a millimetre. These small wave-lengths are conveniently expressed in a unit which is one ten-millionth of a millimetre (10^{-8} cms.) : this is called an Ångström unit (Å.U.). Thus the wave-length of the extreme red rays is about 7600 Å.U., that of the extreme violet about 3800 Å.U.

The wave theory of light appears at first to contradict direct observation. For, as the simplest observations show, light travels from a luminous body only in rays and only in straight lines ; it does not pass round obstacles or bodies which cast shadows. Wave motion behaves quite otherwise; when the waves on a pool meet an obstacle, say a stone which protrudes out of the water, they are seen to draw together again after passing the edges of the stone, so that the water on the far side is also rippled, i.e. the waves have gone round it. We know, too, that sound, which is a wave motion in the air, goes round obstacles ; the sound from a source is heard round corners. But with light this does not seem to be the case. If light waves are allowed to fall on an opaque screen nothing shows behind the screen, so that the light does not seem to go round the corner of an obstacle.

Really it only seems not to do so. As a matter

of fact, light does go round corners as well, and obstacles deflect it from its straight path ; it is just this fact which proves that light must be regarded as a wave motion in spite of appearances to the contrary. In view of the smallness of the wave-length of light, only very small or narrow obstacles must be put in its way if a noticeable deflection of the light from its straight path is to be recognized. In optics this is most perfectly achieved by the use of what

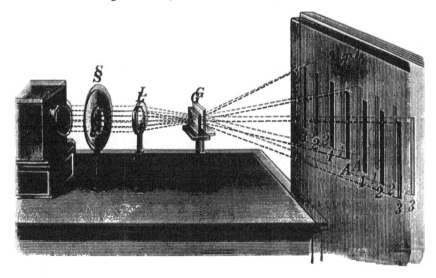

FIG. 24.

is known as a *diffraction grating*. The easiest way to make one of them is to scratch on a glass plate, by means of a very fine dividing engine, a large number of fine lines equidistant from one another. If there are some 100 or 200 lines per millimetre, the transmitted light is seen to be bent far from its straight path. In Fig. 24 a well-known optical experiment is represented, which gives objective demonstration of this deviation or *diffraction of light* as it is called. Light from a source (arc lamp) B,

which has been rendered monochromatic by causing
it to pass through a red glass, goes through a narrow
slit S, and by means of a lens L is thrown on a
board T, where it gives a bright and sharp red image
A of the slit.   But if the diffraction grating is placed,
at G, in the path of the ray of light, there is seen at
once, in addition to the direct image of the slit at A,
a series of lateral images 1, 2, 3, to the right and left
of the central image.   These are the diffracted images,
and obviously they show that the light is not only
propagated in straight lines behind the grating, but
is also deflected sideways.   The reason why separate
bright diffraction images 1, 2, 3 are obtained, instead
of the whole board being more or less illuminated,
lies in the multiplicity of the spacings in the grating :
at each point on the board many separate trains of
waves meet, but, owing to the smallness of the wave-
length, they neutralize one another by interference
at most places, and only reinforce one another at
certain places 1, 2, 3.   The distances of the diffracted
images from one another and from the middle increase with the wave-length of the light used.   Consequently, if a source of light which is polychromatic
is used, the diffraction grating produces automatically
a separation of the different colours or wave-lengths.
The short-waved colours give diffraction images
nearer to, the long-waved farther from, the central
image.   When the white light of the sun or of an
arc lamp is used, the single lines 1, 2, 3, are spread
out into spectra in which all colours from blue to
red merge into one another.

In the theory of optics these diffraction phenomena afford strong evidence for the undulatory

nature of light. At the same time they demonstrate two properties of light which are characteristic of all wave motion, viz. first, the interference of light, i.e. the fact that two light waves may under certain conditions neutralize one another, and second, the deviation from linear propagation, or the " bending round a corner," of light.

In order, then, to demonstrate the undulatory nature of X-rays, it was necessary to show that diffraction effects could also be obtained with them. But even when the finest diffraction gratings that can be made were used, with more than 1700 lines ruled to the millimetre, the X-rays showed no sign of being deviated or diffracted from their straight path. This negative result is, however, no proof that X-rays have not an undulatory nature ; the reason might be that the wave-lengths of X-rays were vastly smaller than those of light. For the grating to be effective, the apertures in it must be so narrow as to include at the most 20 to 50 wave-lengths, and the narrower they are the better. So if the wave-lengths of X-rays were 1000 times less than those of violet light, the grating spaces would be much too large to produce observable diffraction with them : gratings would have to be prepared having not 1000 or 2000 but 1,000,000 lines to the millimetre. But such a task is mechanically impracticable.

A brilliant idea of Prof. v. Laue's led, however, to an unexpected advance which did what was wanted. Instead of attempting the impossible and trying to prepare an artificial grating of the desired and requisite fineness, Laue hit on the idea that gratings of this fineness are, in fact, at our disposal in nature. For in crystals the atoms are arranged like a grating, in that there are, alternately, layers which are occupied by atoms and between them empty spaces ; and the actual distance of these layers from one another is of the order of 8×10^{-8} cms. $= 8$ Å.U., which is just of the order which, as above explained, would be necessary to diffract waves even 1000 times smaller than those of visible light. An

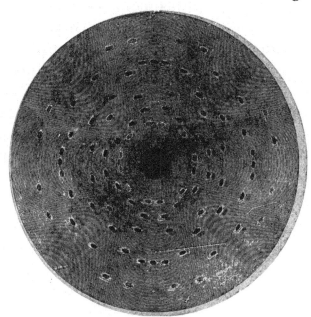

FIG. 25.

experiment with such crystals was completely successful : it was possible, as a matter of fact, to obtain diffraction phenomena when a small pencil of X-rays was allowed to traverse a crystalline plate. A diffraction picture so obtained is shown in Fig. 25 : in this case the X-rays were sent through a crystal of zinc blende and then fell on a photographic plate. The intense black spot at the centre shows where the pencil of rays used came straight through : if the propagation were only linear, this should have been the only one present. But besides this central spot a great number of others are visible, which are symmetrically disposed, and of more or less intensity. These spots are the result of diffraction in the crystal grating, and they exhibit the same symmetry as that of the crystal face used. The crystal of zinc blende has what is called a fourfold axis in the direction in which the X-rays were sent through it. This means that, if the crystal is rotated about this axis, there are four positions, 90° apart, which are precisely equivalent to one another. Similarly, in the diffraction picture each spot is seen to correspond with three others like it, which are displaced by 90° from one another. The symmetry of the crystal is thus reflected in the diffraction picture. This circumstance has caused these X-ray interferences to become an extremely important and successful method in crystallography, by means of which the detailed constitution, or the arrangement of the separate atoms in the crystal, may be investigated. Another consequence of these phenomena was of singular importance in physics, for the wave-lengths of the X-rays used could be determined from the figures produced. The different diffraction spots were shown to be due to wave-lengths of different values present in the non-homogeneous X-radiation used, and calculation indicated that in the X-rays employed there were wave-lengths varying from 0·3 to 1·5 Å.U. The waves are from 600 to 3000 times smaller than those of the most extreme ultra-violet which has yet been produced and measured, in which the wave-length is still 1000 Å.U.

The occurrence of these interference phenomena in the passage of X-rays through crystals may be explained, in a very convenient and suggestive manner, by a method of treatment which was first introduced by Bragg. We have in a crystal a regular network of intersecting planes each of which contains atoms. One series of parallel planes always contains atoms in a certain arrangement, another series of parallel planes which is inclined to the first contains another arrangement of atoms, a third series has yet a different arrangement and so on.

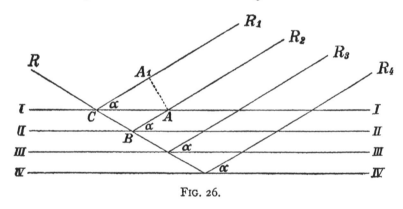

FIG. 26.

We will call each such series of similar parallel planes a *system of lattice planes*. Let us assume that the planes I, II, III, IV in Fig. 26 represent such a system of lattice planes and that a parallel beam of X-rays R_1, R_2, R_3, R_4 of a fixed wave-length λ falls on these lattice planes at practically grazing incidence, i.e. at a very small angle α which is called the *glancing angle*. Then secondary X-rays proceed in all directions from the points of incidence, but in general they neutralize one another by interference. Only when the wave-length of the X-rays bears a certain definite relation to the glancing angle do they not neutralize but reinforce one another : in that case a beam of X-rays leaves the crystal in the direction R, that is, in the same direction as if the beam of rays R_1 to R_4 had been directly reflected from the lattice planes. This reinforcement occurs when the difference between the routes of two successive rays (the path-difference) amounts to a whole wave-length (or to two or three whole wave-lengths). The path-difference between the rays R_2 and R_1 is equal to $AB + BC - A_1C$; this must be equal to a wave-length. (The resulting relation between the wave-length λ, the glancing angle a, and the distance d between the parallel lattice planes is simply $\lambda = 2d \sin a$.) But hence it also follows that if the parallel beam of rays, incident at the angle a, has not only one single wave-length, i.e. is not homogeneous but is a mixture of several wave-lengths, then only one ray of the beam issues in the direction R, namely, that which actually has the particular wave-length λ. If the incident beam does not contain this wave-length, then nothing will be reflected from these lattice planes at this glancing angle. If this wave-length is present in the beam, only the ray with this wave-length will be reflected. But this means that at this angle the crystal separates and reflects one particular wave-length from a mixture of wave-lengths, i.e. that it produces from the non-homogeneous X-rays a homogeneous or *monochromatic* radiation, as it is called, in the direction R.

What holds for one glancing angle a, that it separates and reflects from the non-homogeneous rays those of wave-length λ, holds also for another glancing angle a^1 and another wave-length λ^1. Hence

7

it follows that if a beam of X-rays of different wave-lengths, which is not parallel but divergent, strikes such a crystal then the crystal reflects the separate wave-lengths in different directions, i.e. it forms by reflection a fan of rays, each of which has a certain definite wave-length. For the non-parallel, divergent rays make various glancing angles with the lattice planes, and at each of these angles a ray of a different wave-length is reflected. The crystal thus decomposes the incident, non-homogeneous beam of rays into a spectrum of X-rays so that only one ray, with an absolutely definite wave-length, is reflected in each direction. Just as in optics we decompose white light by a prism into a spectrum in which each colour, or light of different wave-length, is refracted at a different angle so that the colours are separated, so a crystal decomposes a non-homogeneous X-radiation of different wave-lengths into an X-ray spectrum in which the individual rays of different wave-lengths are separated by reflection.

We have thus gained a method of *X-ray spectroscopy*, a method which allows us not only to separate X-rays of different wave-lengths, but also to measure these wave-lengths directly. It is only necessary to measure the glancing angle for each ray with a particular crystal, in which the distance of the lattice planes is known from crystallographic considerations, in order to obtain thence the wave-length.

Measurements of this sort were first carried out after Bragg by the young English physicist Moseley, who fell in Gallipoli, sacrificed to the world war. It was found in these investigations that every X-ray bulb emits, as we have mentioned, a non-homogeneous, mixed radiation. In fact a spectrum is obtained which comprises all possible wave-lengths within certain limits, corresponding thus with the spectrum of a white body. But, in addition, one finds in the spectrum very strong isolated lines, i.e. certain definite wave-lengths. These definite and strong wave-lengths depend on the nature of the metal of which the anticathode of the bulb is made. They are called the *characteristic radiation* of the anticathode metal. It is, in particular, these sharp lines of the characteristic radiation which are of pre-eminent importance for our study.

In carrying out these measurements it has been found advantageous not to fix the crystal rigidly, since then irregularities in the reflecting faces have a detrimental influence, but rather to rotate it slowly so that the disturbing effects of the faces cancel out on the average. The reflected rays fall on a photographic plate contained in a box. An arrangement of apparatus as constructed by Prof. Siegbahn of Lund, to whom is due the most accurate and complete investigation of these spectra, is shown in Fig. 27. On a marble plate stands a pillar carrying two diaphragms B_1 and B_2 of lead, the slit-shaped aperture in which may be made smaller or larger. The X-rays pass through these slits to the crystal plate K, which is slowly rotated backwards and forwards by the clockwork U, arrangements for exact adjustment being provided. The reflected spectrum falls on a box P, which contains the photographic plate. The spectra so formed are exemplified by Fig. 28, in which a series of sharp lines is seen against the uniformly darkened background ; these

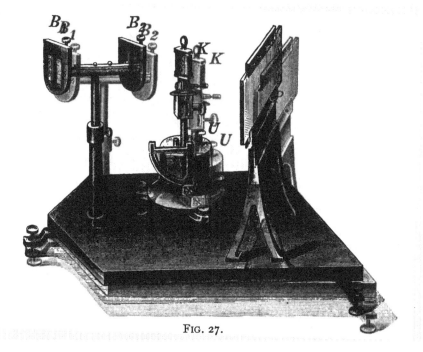

FIG. 27.

are the characteristic lines, in the present case from a platinum anticathode.

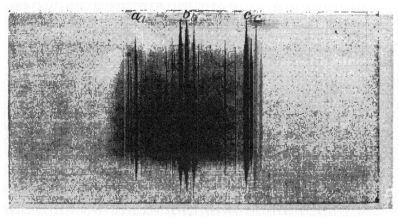

FIG. 28.

Every metal, when used as anticathode in a bulb, shows similar lines of definite wave-length. Experiment has shown that the wave-lengths of these characteristic lines are the shorter, the higher the atomic weight of the metal concerned is. It has further been demonstrated that the lines for the lighter elements from sodium to neodymium have all the same structure. They are called the K-lines or the K-spectrum. They consist always of five lines which are distinguished as $K\alpha_1$, $K\alpha_2$, $K\beta_1$, $K\beta_2$ and $K\gamma$; of these the lines $K\alpha_1$ and $K\alpha_2$, which are close together, are the strongest. The β-lines similarly form a doublet, and in some substances the γ-line is also double. These lines are present with all the substances mentioned. If the cathode consists of an alloy, e.g. brass, the lines of both constituents, copper and zinc, appear together on the plate. Even if the anticathode is simply covered with the salt of a metal the lines of the constituents are shown, e.g. potassium chloride gives the lines of chlorine and of potassium.

With the heavier of these elements, from zinc onwards, a second system of lines is found, distinct from the first, with wave-lengths considerably higher than those of the K-lines. This is known as the L-spectrum. It consists of from ten to fourteen lines, the strongest of which is called the $L\alpha_1$-line. Here again the wave-length of the lines diminishes with increasing atomic weight. Among the heaviest elements, from neodymium up to uranium, only the L-series has hitherto been found and not the K-series. On the other hand, for the heaviest elements of all, from dysprosium to uranium, a spectrum has been found with still longer waves; this is the M-spectrum, and consists of from three to five lines, some of them doublets.

The wave-lengths of the $K\alpha_1$-lines for a few elements are given in the following table, in which the atomic weight and the number of the element in the periodic system or the *atomic number*, are also stated.

Atomic Number	Element.	Atomic Weight.	Wave-length of $K\alpha_1$ in 10^{-8} cms.
11	Sodium	23·00	11·951
15	Phosphorus	31·04	6·168
20	Calcium	40·07	3·355
25	Manganese	54·93	2·093
30	Zinc	65·37	1·433
35	Bromine	79·92	1·035
40	Zirconium	90·6	0·788
45	Rhodium	102·9	0·615
50	Tin	118·7	0·487
55	Cæsium	132·81	0·398
60	Neodymium	144·3	0·330

It is better to use, instead of the wave-length, the *frequency of vibration*, which is higher the smaller the wave-length is. The frequency of vibration (i.e. the number of vibrations per second) is obtained from the wave-length by dividing the latter into the velocity of propagation of motion in the ether, which is 3×10^{10} cm./sec. As the atomic weight increases the wave-lengths of the K-lines continually diminish, so that the vibration frequencies of the K-vibrations increase ; it is found, in fact, that the square roots of the vibration frequencies increase roughly as the atomic weights.

It was first recognized by Moseley that there is a very simple and direct relationship not between the square root of the vibration frequency and the atomic weight, but between the square root of the vibration frequency and the atomic number of the element.

In the subjoined table are given the vibration frequencies ν for the above substances together with the square roots of the vibration frequencies.

$K\alpha_1$–radiation.

Atomic Number.	Element.	Atomic Weight.	Vibration Frequency ν of $K\alpha_1$.	Square Root of Vibration Frequency.
11	Sodium	23·00	$0\cdot2510 \times 10^{18}$	$5\cdot010 \times 10^8$
15	Phosphorus	31·04	0·4863 ,,	6·973 ,,
20	Calcium	40·07	0·8942 ,,	9·456 ,,
25	Manganese	54·93	1·433 ,,	11·97 ,,
30	Zinc	65·37	2·088 ,,	14·44 ,,
35	Bromine	79·92	2·898 ,,	17·03 ,,
40	Zirconium	90·6	3·808 ,,	19·51 ,,
45	Rhodium	102·9	4·878 ,,	22·08 ,,
50	Tin	118·7	6·160 ,,	24·82 ,,
55	Cæsium	123·81	7·537 ,,	27·45 ,,
60	Neodymium	144·3	9·090 ,,	30·15 ,,

The simple dependence of the vibration frequency on the atomic number is immediately obvious from Fig. 29. For if the atomic numbers are plotted horizontally and the square roots of the vibration frequencies vertically, all the points so obtained lie exactly and definitely on a straight line. This means that *the square root of the vibration frequency is linearly connected with the atomic number of the element.*

If the relation between the square root of the vibration frequency and the atomic weight is similarly plotted the curve of Fig. 30 results; this has in general the same course but exhibits at various points marked deviations from the straight line, deviations which are greater than correspond with the accuracy of the measurements. Even without the curves the same thing can be seen from the above numbers. For if we take from the foregoing table

(Square root of Frequency) × 10⁻⁸
30
25
20
15
10
5
0
5
10
15
20
25
30
35
40
45
50
55
60
Ka1
Atomic number.

FIG. 29.

(Square root of Frequency) × 10⁻⁸
30
25
20
15
10
5
10
20
30
40
50
60
70
80
90
100
110
120
130
140
150
Atomic weight.

FIG. 30.

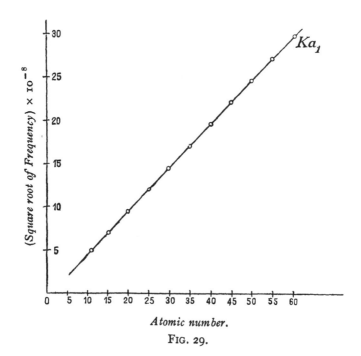

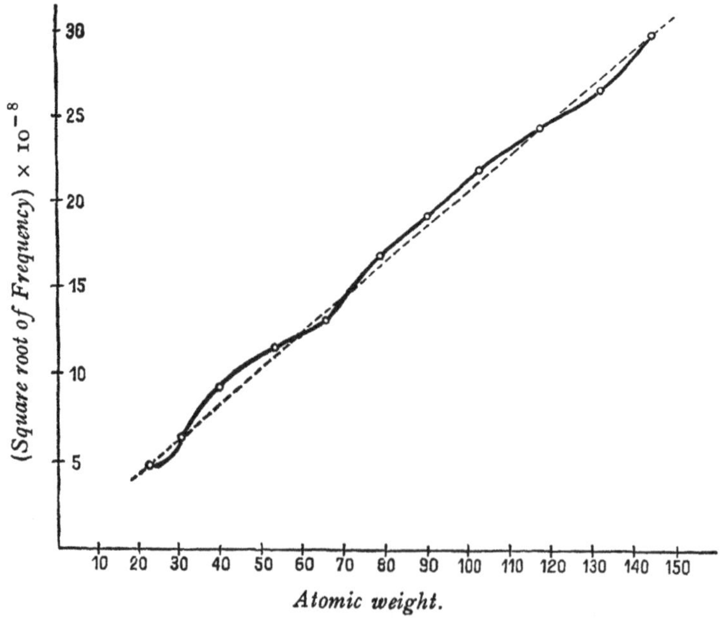

quotients of the differences according to the scheme, e.g.

$$\frac{\text{Cæsium—phosphorus,}}{\text{Bromine—phosphorus}}$$

we obtain for the square roots of the vibration frequencies

$$\frac{20 \cdot 477}{10 \cdot 057} = 2 \cdot 036,$$

for the atomic numbers

$$\frac{40}{20} = 2 \cdot 000,$$

and for the atomic weights

$$\frac{101 \cdot 77}{48 \cdot 88} = 2 \cdot 082.$$

The first and second quotients agree better than the first and third. The difference is not great but is clearly present.

Hence we draw the conclusion that the square roots of the vibration frequencies of X-rays depend directly and simply on the atomic numbers of the elements, not on their atomic weights.

This law holds not only for the K-radiation, but also for the softer L-radiation, which could be traced up to the elements of the highest atomic weight. In the following table are given similarly for a number of elements their atomic numbers, atomic weights, wave-lengths of the $L\alpha_1$-line, their vibration frequencies, and the square roots of the latter.

$L\alpha_1$–radiation.

Atomic Number.	Element.	Wave-length of $L\alpha_1$.	Vibration Frequency.	Square Root of Vibration Frequency.
30	Zinc	$12\cdot222 \times 10^{-8}$	$0\cdot2454 \times 10^{18}$	$4\cdot954 \times 10^{8}$
35	Bromine	$8\cdot357$,,	$0\cdot3590$,,	$5\cdot991$,,
40	Zirconium	$6\cdot083$,,	$0\cdot4932$,,	$7\cdot023$,,
45	Rhodium	$4\cdot596$,,	$0\cdot6527$,,	$8\cdot079$,,
50	Tin	$3\cdot592$,,	$0\cdot8352$,,	$9\cdot139$,,
55	Cæsium	$2\cdot886$,,	$1\cdot039$,,	$10\cdot20$,,
60	Neodymium	$2\cdot365$,,	$1\cdot269$,,	$11\cdot26$,,
74	Tungsten	$1\cdot471$,,	$2\cdot040$,,	$14\cdot28$,,
80	Mercury	$1\cdot240$,,	$2\cdot420$,,	$15\cdot55$,,
82	Lead	$1\cdot175$,,	$2\cdot554$,,	$15\cdot98$,,
88	Radium	$1\cdot010$,,	$2\cdot970$,,	$17\cdot23$,,
90	Thorium	$0\cdot957$,,	$3\cdot135$,,	$17\cdot71$,,
92	Uranium	$0\cdot911$,,	$3\cdot293$,,	$18\cdot14$,,

From these numbers also we may similarly recognize a smooth linear relation between the square roots of the vibration frequencies and the atomic numbers, not the atomic weights. If, for example, we take the quotients of differences according to the scheme.

$$\frac{\text{Thorium—zinc}}{\text{Neodymium—zinc}}$$

we have for the square roots of the vibration frequencies

$$\frac{12\cdot76}{6\cdot31} = 2\cdot022,$$

for the atomic numbers

$$\frac{60}{30} = 2\cdot000,$$

and for the atomic weights

$$\frac{166\cdot78}{78\cdot93} = 2\cdot113.$$

The deviation of the first two quotients from one another is only 1 per cent. ; between the first and third it amounts to 4·5 per cent. This shows again, and still more sharply in the present case, that the square roots of the vibration frequencies depend in a linear manner on the atomic numbers.

The same thing holds, too, for the still softer M-rays, for which the corresponding numbers for the Mα-line (according to Stenström) are given for a number of elements.

Mα–radiation.

Atomic. Number	Element.	Atomic Weight.	Wave-length. of Mα.	Vibration Frequency.	Square Root of Vibration Frequency.
66	Dysprosium	162·5	$9·509 \times 10^{-8}$	$0·3155 \times 10^{18}$	$5·617 \times 10^{8}$
70	Aldebaranium	173·8	8·123 ,,	0·3693 ,,	6·077 ,,
74	Tungsten	184·0	6·973 ,,	0·4302 ,,	6·559 ,,
79	Gold	197·2	5·819 ,,	0·5156 ,,	7·180 ,,
82	Lead	207·1	5·275 ,,	0·5687 ,,	7·541 ,,
90	Thorium	232·4	4·129 ,,	0·7266 ,,	8·525 ,,
92	Uranium	238·5	3·901 ,,	0·7690 ,,	8·769 ,,

The ratio $\dfrac{\text{uranium—dysprosium}}{\text{gold—dysprosium}}$

gives for the square roots of the vibration frequencies

$$\frac{3·152}{1·563} = 2·017,$$

for the atomic numbers

$$\frac{26}{13} = 2·000,$$

and for the atomic weights

$$\frac{76·0}{34·7} = 2·191,$$

so that here again it is the atomic numbers and not the atomic weights which are of importance.

The law thus enunciated is known as *Moseley's*

law of high-frequency spectra. The X-ray spectra
are called high frequency spectra because of their
very high vibration frequencies. Moseley's law
states that the square roots of the vibration frequencies are linear functions of the atomic numbers
of the elements. It is expressed by the simple
equation

$$\sqrt{\nu} = c\,(N - a)$$

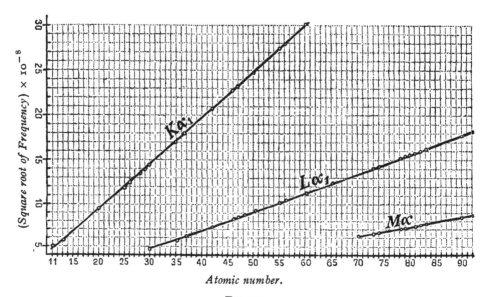

Atomic number.

FIG. 31.

where ν is the vibration frequency and N the atomic
number of the element, and a and c are constants.
The constants c and a for the K-radiation are different
from those for the L-radiation and for the M-radiation.

This law is clearly shown for the K-, L- and
M-radiations by Fig. 31 in which it is seen that the
relation between the square root of the vibration
frequencies on the one hand and the atomic numbers
on the other is accurately represented, for all three
series, by straight lines.

Now these results are supplementary but very convincing evidence for Rutherford's nuclear theory of the atom. The vibrations of X-rays depend essentially on the atomic number of the elements, i.e. of the nuclei, so that the vibration frequencies increase regularly with the nuclear charge.

There is no hint of a periodic variation of the vibration frequencies with increasing nuclear charge, such as there is for the chemical behaviour of the elements. It follows, therefore, that the periodicity occurring in chemistry, which has found its expression in the periodic system of the elements, cannot depend on the nuclear charge, but only on the arrangement of the outer electrons, with which we have not hitherto dealt.

But we have now obtained, from this simple regular relation of the high-frequency vibrations, an unexceptionable method, firstly, for deciding on the actual serial order of the elements in the periodic system, in any cases of doubt, secondly, for stating definitely where there are still gaps corresponding with undiscovered elements and, finally, for filling up correctly the particularly large gap which hitherto extended between cerium and tantalum.

First as concerns the serial order of the elements, several cases have long been known where the arrangement according to atomic weights did not give the right position. The three elements, iron, cobalt, and nickel, arranged themselves, according to their chemical behaviour, in the order given ; but their atomic weights are respectively 55·84, 58·97, and 58·68, according to which nickel should come before cobalt. Measurement of the $K\alpha_1$ wave-lengths resulted in the values 1·928 Å.U. for iron, 1·781 for cobalt, and 1·653 for nickel, whence the square roots of the vibration frequencies are 12·47, 12·98, and 13·47 × 10⁸ ; it follows without a doubt that the nuclear charge of nickel is higher by one than that of cobalt, so that the proper serial order of the elements in the system is here not the order of the atomic weights.

Similarly, it has long been known that tellurium with atomic weight 127·5 belongs to the sixth column of the system (with sulphur and selenium), while iodine with the lower atomic weight 126·92 belongs to the seventh column (with chlorine and bromine), thus coming after tellurium in spite of having a lower atomic weight. X-ray spectra give, for the wavelengths of the K_{α_1}-lines, 0·456 Å.U. for tellurium and 0·437 for iodine, the square roots of the vibration frequencies being thus 25·65 × 10⁸ and 26·20 × 10⁸ ; this demonstrates that here again the correct order does not agree with the order of the atomic weights. Similarly argon (atomic weight 39·88) undoubtedly precedes potassium (atomic weight 39·10).

The largest lacuna in the periodic system was for many years that between cerium with atomic weight 140·25 and tantalum with atomic weight 181·5 ; in this space fell the rare earths, but it was not known how many elements were required to fill it. Investigation of the L-spectra gave at once an answer to the latter question. For, according to the linear relation, cerium certainly corresponded with the atomic number 58 and tantalum equally certainly with the atomic number 73, so that there should be fourteen other elements between them.

Of these praseodymium, neodymium, samarium, terbium, holmium, erbium, thulium, and gadolinium had long been known, others (europium, dysprosium, aldebaranium and cassiopeium) were found later and two are still unknown.[1]

It has been established by these investigations that from hydrogen to uranium there are ninety-two different elements and no more, if we except isotopes (p. 83), of which a number of elements are composed but which have, however, all the same nuclear charge and consequently occupy the same position in the periodic system. In arranging the elements in the periodic system, difficulty is encountered with the rare earths. They are a class by themselves and do not conform with the regularity otherwise exhibited in the periodic system. The arrangement on page 112 (after Fajans) is accordingly given for the periodic system with nine columns of which columns O and VIII are single, the others double ; in front of each element is placed its ordinal number, and below it its atomic weight (O = 16). Besides two unknown rare earths, numbers 61 and 72[1], there are only four substances, 43, 75, 85, and 87 unknown. The radio-active substances, 84 polonium, 86 radium emanation, 89 actinium, 91 proto-actinium, are conjecturally inserted. Those elements of which isotopes have been recognized are distinguished by an asterisk.

Since the nucleus theory of the atom has thus obtained very strong support from the observation

[1] Urbain assigns the atomic number 72 to a rare earth element " celtium," but Coster claims this number for a new member of the zirconium group which he calls " hafnium."—Tr.

O.	I. a	I. b	II. a	II. b	III. a	III. b	IV. a	IV. b	V. a	V. b	VI. a	VI. b	VII. a	VII. b	VIII.
1 H 1·008															
2 He 4·00	3 Li* 6·92		4 Be 9·1		5 B* 11·0		6 C 12·00		7 N 14·01		8 O 16·00		9 F 19·0		
10 Ne* 20·2	11 Na 23·0		12 Mg* 24·32		13 Al 27·1		14 Si* 28·3		15 P 31·04		16 S 32·06		17 Cl* 35·46		
18 A* 39·88	19 K* 39·10		20 Ca 40·07		21 Sc 45·1		22 Ti 48·1		23 V 51·0		24 Cr 52·0		25 Mn 54·93		26 Fe 55·84 27 Co 58·97 28 Ni* 58·68
		29 Cu 63·57		30 Zn* 65·37		31 Ga 69·9		32 Ge 72·5		33 As 74·96		34 Se* 79·2		35 Br* 79·92	
36 Kr* 82·92	37 Rb* 85·45		38 Sr 87·63		39 Y 88·7		40 Zr 90·6		41 Nb 92·5		42 Mo 96·0		43 —		44 Ru 101·7 45 Rh 102·9 46 Pd 106·7
		47 Ag 107·88		48 Cd 112·40		49 In 114·8		50 Sn* 118·7		51 Sb* 120·2		52 Te 127·5		53 I 126·92	
54 Xe* 130·2	55 Cs 132·81		56 Ba 137·37		57 La 139·0		58 Ce 140·25	59 Pr 140·9	60 Nd 144·3	61 —	62 Sa 150·4	63 Eu 152·0	64 Gd 157·3		
							65 Tb 159·2	66 Dy 162·5	67 Ho 163·5	68 Er 167·7	69 Tm 168·5	70 Ad 173·5	71 Cp 175·0		
							72 —		73 Ta 181·5		74 W 184·0		75 —		76 Os 190·9 77 Ir 193·1 78 Pt 195·2
		79 Au 197·2		80 Hg* 200·6		81 Tl 204·4		82 Pb* 207·20		83 Bi* 209·03		84 Po (210·0)	85 —		
86 Ra Em* (222·0)	87 —		88 Ra 226·0		89 Ac (227)		90 Th 232·15		91 Pa (230)		92 U 238·2				

of high-frequency spectra, we may go on, with considerable confidence, to examine further deductions from it.

For an atom to be neutral or non-electric, as is in general the case, each one must contain just as many free negative electrons outside the nucleus as there are positive elementary charges in the nucleus. We should therefore expect that a neutral atom of hydrogen should have on the outside of it one electron, that of helium two, and finally that of uranium ninety-two negative electrons. But an atom may also have attached to it either fewer or more of such negative electrons. In the former case it is a positive atomic ion with one or more free positive charges, in the latter case it is a negative ion with one or more free negative charges. From the fact that atoms have a certain chemical *valency*, being mono-, di-, tri-, tetra- or poly-valent, we must conclude that some of these electrons on the outside behave differently from others. As a rule, most of them are, under normal circumstances, fairly closely combined with the nucleus ; but some of them, one in a mono-valent substance, two in a di-valent, three in a tri-valent and so on, are more loosely attached to the nucleus, so that they can be more readily separated from or added to it. For it is by means of these valency electrons that atoms combine to form molecules. Since a copper ion, e.g., is positively di-valent, we must assume that it does not contain the twenty-nine negative electrons which the atom in the neutral state would have, but only twenty-seven, and that it is the two missing electrons that determine its di-valency. The negatively di-valent

8

oxygen ion, on the other hand, can easily bind to itself two more negative electrons in addition to the eight electrons which the neutral oxygen atom has. But as to the manner and method in which the negative electrons which lie or revolve about the nucleus are arranged on the outside of it, no information is immediately available from the facts hitherto adduced : some of these facts though, in particular Moseley's law, demand further explanation on this constitution of the atom.

As to the positive charges on the nucleus of the atom, a number of different opinions may be held. The positive charge on the nucleus of the hydrogen atom is a small volume (of radius 10^{-16} cms.), which contains one elementary charge. Rutherford is of opinion that this hydrogen nucleus is the positive electron itself ; this has much in its favour. He accordingly proposes a special name *proton* for this hydrogen nucleus. Thus by electron would always be understood the negative elementary quantum, as is already customary, by proton the positive quantum. The nuclear charges of all other elements would then be composed of these positive hydrogen nuclei or protons together with the requisite negative nuclear electrons. This is, in its essentials, Prout's hypothesis (p. 6). The objection which we raised against the latter, that, according to it the atomic weights of all substances must be whole numbers, which they are not, is now no longer valid owing to the discovery of isotopes of many elements (p. 88). Small deviations from whole numbers do, however, certainly occur. As already explained above, the atomic weight of hydrogen, if that of oxygen is taken as 16, is not 1 but 1·008 and similar small deviations from whole numbers are established for many elements. But, according to Prout's theory as now interpreted, there can be no necessity for atomic weights being exactly whole numbers. For the atomic weight is composed of the masses of the nucleus and of the negative electrons, and, moreover, all these masses are apparent. Now even though the apparent mass of a negative electron is only 1/1835 of the mass of a hydrogen atom, it is to be expected that if many of these negative electrons are present in and outside the nucleus, small deviations from whole numbers will occur for the atomic weights. Another point follows. If the negative electrons outside the nucleus are in motion and their velocity happens, in some particular atom, to be very large, i.e. comparable with the velocity of light, then their apparent mass would also be greater, and under these circumstances even considerable deviations of the atomic weights from whole numbers could occur. Thus the objection to Prout's hypothesis, which was rightly raised at an earlier stage, has here no longer any force, for the number of negative electrons in and outside the nucleus may in some circumstances be considerable. Again, the mass of a hydrogen nucleus depends to some extent, since it is apparent mass (p. 76), on whether other positive or negative charges are in its immediate neighbourhood ; it may accordingly be either greater or smaller. This is another reason for possible small deviations of the atomic weights from whole numbers either upwards or downwards. In short, there is no longer any valid objection to the supposition that all atoms are composed of protons and electrons.

On the other hand, the assumption that the hydrogen nucleus is itself the positive electron and that there is only one kind of positive electron, is founded, in the first place, on nothing more than our desire for simplicity. If it is the case, the helium nucleus, e.g., with atomic weight 4 and with two positive charges, which is the α-particle, would have to be composed of four hydrogen nuclei and two negative electrons. Its radius would then no longer be of the order 10^{-16} cms. but, by reason of the size of the negative electrons, of the order of 10^{-13} cms. Nuclei, other than that of hydrogen, would then occupy a volume at least as great as the negative electron itself. That hydrogen nuclei are in fact present in other nuclei and that the helium nucleus is of the size last mentioned is demonstrated by the decomposition of the nitrogen atom, discovered by Rutherford, which is discussed in the sixth lecture.

LECTURE V

LINE SPECTRA AND BOHR'S MODEL OF THE ATOM

PRIOR to the discovery of radio-activity, the most sensitive method available in chemistry for the detection of the smallest traces of substances was, as is well known, that of spectroscopic analysis ; and it was by this method that the discovery of a large number of new elements was rendered possible. The principle of it is to cause the light emitted by incandescent vapours or by luminous gases to pass through a slit and a prism, whereby it is spread out into a spectrum. What then appears is not, as with white-hot solids, an extended band which contains all the colours from red to violet merging continuously into one another, i.e. a complete spectrum, but only a few or sometimes many distinct coloured *lines ;* this is what is called a *line spectrum.*

The lines which are produced under these conditions are so characteristic of the chemical elements that are present in the luminous vapour that it is possible to deduce, inversely, the presence of a particular element from the occurrence of certain lines. In the most accurate experiments a *diffraction grating* is used instead of a prism for the production and examination of the spectrum ; the best gratings are those of Rowland, consisting of a plate of speculum metal on which are ruled 1700 fine equidistant lines per millimetre. If the light from a luminous vapour or gas is allowed to pass through a slit and fall on such a grating, diffraction spectra are obtained by reflection on both sides of the direct image of the slit ; there are a primary, a secondary, and a tertiary spectrum at increasing angular distances from the direct image. From the accurately-measured angular distance of a line in one of these diffraction spectra from the central image of the slit and the number of lines per millimetre on the grating it is possible to determine with great accuracy the wave-length of the particular coloured line : this is, in fact, the principal method for such measurements.

The wave-lengths of all colours are very small. In what follows they will always be expressed in Ångström units (Å.U.) ; this unit, as above explained (p. 90) is equal to one ten-millionth of a millimetre i.e. 10^{-8} cms.

Certain of the lines shown by a luminous gas are always distinguished by special brightness and sharpness. These are usually regarded as *the* lines of the element concerned. But careful examination with an efficient diffraction grating (a grating with many lines, i.e. of high dispersion) shows a further large number of other and weaker lines ; it often shows, too, that certain lines, which had previously been regarded as single, consist of two or more lines very close together. The appearance of the spectrum also depends considerably on the temperature of the luminous vapour or on the type of electrical discharge which is used to excite luminosity in Geissler tubes.

The visible spectrum, which extends from the red with a limiting wave-length of about 7600 Å.U. to the violet with a limit of about 3800 Å.U. has, as is well known, continuations in both directions; these are not, it is true, directly recognizable by the eye, but with the aid of physical instruments they can be just as accurately investigated as the visible spectrum. The continuation beyond the red is known as the *infra-red*. The lines occurring here, whose position can be fixed by their heating effect on suitable apparatus, have wave-lengths greater than 7600 Å.U. They may be traced easily up to 50,000 Å.U. and with sufficiently delicate apparatus

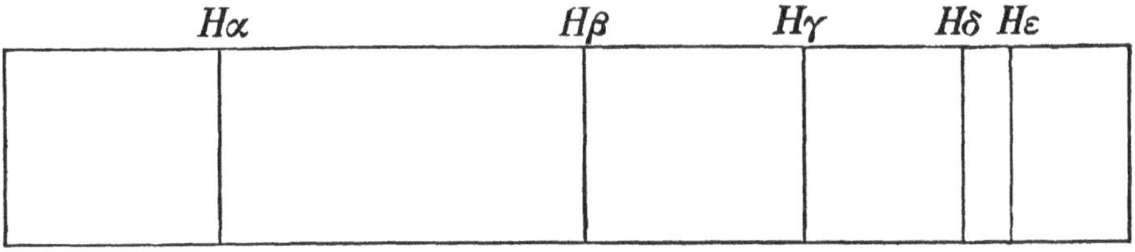

$H\alpha$		$H\beta$	$H\gamma$	$H\delta$ $H\varepsilon$

FIG. 32.

very much farther still, even up to $\frac{1}{3}$ mm., i.e. over three million Å.U. The invisible part of the spectrum beyond the violet end, the *ultra-violet*, in which most substances give very many lines, is easily investigated by photography; for the ultra-violet rays for a long way down act on a photographic plate. In this manner lines may readily be found with wave-lengths down to 1800 Å.U. and with suitable apparatus as far as 1000 Å.U.

As an example of line spectra, the spectrum of the principal lines of hydrogen is given in Fig. 32, as seen when a Geissler tube (spectrum tube) filled with hydrogen is observed with a spectroscope (slit and prism). In this case four lines are usually seen, and at suitable exhaustion the fifth as well ; these five lines are called Hα, Hβ, Hγ, Hδ, Hε, and they have the following wave-lengths :—

	Colour.	Wave-length in Å.U.
Hα	red	6563
Hβ	green	4861
Hγ	dark blue	4341
Hδ	violet	4102
Hε	violet	3970

A second example is afforded by the spectrum of the principal lines of helium in Fig. 33.   Here seven

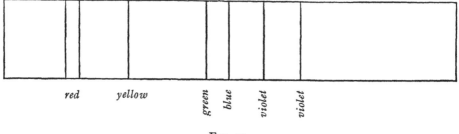

FIG. 33.

bright lines are visible which have the following colours and wave-lengths :—

Colour.	Wave-length in Å.U.
red	7065
red	6678
yellow	5876
green	5016
blue	4923
violet	4713
violet	4471

Similarly the different elements show a varying number of lines in their spectra. The line spectra of the alkali metals (group I*a* of the periodic system), are also relatively simple, those of the alkaline earths (group II*a* of the periodic system) are rather more complicated, but with the remaining elements the spectra are extraordinarily rich in lines.

Since the spectra have a special composition for each chemical element, they obviously depend on the atoms, and since each line corresponds with a definite vibration frequency, it is evident that definite periodic movements are taking place within the atom. Thus the deduction could have been, and was, made some time ago from the existence of spectra, that atoms are not merely spaces filled with mass, but that they consist of parts which can execute periodic motions of different kinds.

In acoustics, too, a body can execute vibrations of very different kinds ; but though the attempt might at first be made to follow out an analogy from the well-known phenomena of sound, it is soon seen that it leads in this case to no result. When a column of air in a pipe is caused to oscillate or a stretched piano-string or violin-string to vibrate, it is certainly found that the air column or the string does not execute only one vibration of a single definite frequency, but that it begins simultaneously all sorts of vibrations of very different frequencies. Such a column of air or a string gives not a *simple tone* but a *note*, which in acoustics signifies a combination of different simple tones. But if such a note is analysed to find the separate tones of which the note is composed, which can easily be done by Helmholtz's method, the following is what is found. Of all these tones one has the smallest frequency ; this is known as the *fundamental* of the note. In most cases it is also of the greatest intensity and determines the musical " height " of the note. Now all the other tones contained in the note have always frequencies which are simply related to that of the fundamental. If the frequency of the fundamental is denoted by 1 the other tones of the note have some of the frequencies of the series 2, 3, 4, 5, 6, 7, 8. . . . Their frequencies are thus integral multiples of that of the fundamental. This series of tones associated with a fundamental are called its *overtones*. All the overtones may be present—with varying intensity—in a note or only the odd or only the even numbered ones, or else some consecutive overtones may be missing from the series and so on ; in short, the combination of overtones with the fundamental may be very varied. On this depends the variation in *timbre* of the note which is given by different instruments for the same fundamental. What we are here concerned with is that the overtones always have vibration frequencies which are whole multiples of the fundamental. Hence it follows that, counting up from the fundamental, there is in the first octave no overtone, in the second there are two (2, 3), in the third octave four (4, 5, 6, 7), in the fourth eight (8, 9, 10, 11, 12, 13, 14, 15) and so on. The difference in frequency between successive overtones in any octave is always the same, namely 1, being equal to the frequency of the fundamental.

If this were also the case with the vibrations which give rise to the line spectra, the separate lines of a diffraction spectrum, expressed as frequencies, would all be equidistant from one another, except for some which might be missing from the series. But this is far from being the case, so that analogy with sound vibrations leads to no result in the investigation of line spectra.

A law governing the arrangement of lines in a spectrum was first discovered in the case of hydrogen ; this was found purely empirically, simply by trial and error. It was a Swiss, Balmer, who recognized (1885) a very striking regularity in this case. The above-mentioned wave-lengths of the five hydrogen lines Hα to Hε may be represented by the following numbers :—

Name.	Wave-length observed.	Represented by.	Wave-length calculated.
Hα	6563	$3646 \cdot 13 \times \dfrac{9}{9-4}$	6563
Hβ	4861	$3646 \cdot 13 \times \dfrac{16}{16-4}$	4861
Hγ	4341	$3646 \cdot 13 \times \dfrac{25}{25-4}$	4341
Hδ	4102	$3646 \cdot 13 \times \dfrac{36}{36-4}$	4102
Hε	3970	$3646 \cdot 13 \times \dfrac{49}{49-4}$	3970

Thus the factor, by which the number 3646·13 must be multiplied in order to give the observed values, has, for the successive lines, the magnitudes

$$\frac{3^2}{3^2-2^2}, \quad \frac{4^2}{4^2-2^2}, \quad \frac{5^2}{5^2-2^2}, \quad \frac{6^2}{6^2-2^2}, \quad \frac{7^2}{7^2-2^2}.$$

It was not only for the five lines in the visible spectrum that this striking law proved to be valid.

In the ultra-violet yet another series of hydrogen lines has been discovered, some of them actually in Geissler tubes, some of them in the spectrum of the solar prominences, and all twenty-four of these other lines could be reproduced by the same law, the numerator having successively the values 8^2, 9^2 . . . to 31^2, the denominator the values $8^2 - 2^2$, $9^2 - 2^2$, and so on to $31^2 - 2^2$. If m stand for one of the numbers from 3 to 31, the wave-length λ of the corresponding hydrogen line (in Å.U.) is determined by

$$\lambda = 3646 \cdot 13 \times \frac{m^2}{m^2 - 2^2}$$

This extremely surprising formula is called *Balmer's formula*. The twenty-nine lines of hydrogen which are embraced by it are grouped together as a *series* of hydrogen.

It is preferable to introduce, instead of the wave-lengths, the vibration frequencies of these lines. The product of wave-length and vibration frequency is equal to the velocity of propagation of light, i.e. to 3×10^{10} cm./sec. $= 3 \times 10^{18}$ Å.U./sec. ; accordingly the vibration frequency

$$= \frac{3 \times 10^{18}}{\text{wave-length}}$$

Thus for the mth line of hydrogen

Vibration frequency

$$= \frac{3 \times 10^{18}}{3646 \cdot 13} \left(1 - \frac{2^2}{m^2} \right)$$

$$= \frac{3 \times 10^{18} \times 4}{3646 \cdot 13} \left(\frac{1}{2^2} - \frac{1}{m^2} \right)$$

$$= 3 \cdot 291 \times 10^{15} \left(\frac{1}{2^2} - \frac{1}{m^2} \right).$$

The vibration frequencies of all these twenty-nine lines are thus reproduced by this law when the number m is made to assume *seriatim* all integral values from 3 to 31.

It is interesting to note that two hydrogen lines have been found in the infra-red, whose vibration frequencies can be represented in a very similar manner by the formulæ $3\cdot291 \times 10^{15} \left(\frac{1}{3^2} - \frac{1}{4^2} \right)$ and $3\cdot291 \times 10^{15} \left(\frac{1}{3^2} - \frac{1}{5^2} \right)$. The corresponding wave-lengths observed were

$$18751 \text{ Å.U. and } 12818 \text{ Å.U.}$$

and those calculated from the above formulæ are

$$18752 \text{ Å.U. and } 12819 \text{ Å.U.}$$

Thus with them the difference which occurs is not

$\frac{1}{2^2} - \frac{1}{m^2}$ but $\frac{1}{3^2} - \frac{1}{m^2}$ so that they do not belong to

the original Balmer's series but form a series with an exactly similar law.

The factor $3\cdot291 \times 10^{15}$ which in the first place occurred only for the hydrogen spectrum, is, as was first recognized by Rydberg, of general significance for all spectra. This factor is known as *Rydberg's constant* and is abbreviated as the letter R.

The excellence of the agreement of Balmer's formula, and of the correspondingly constructed one just mentioned, with the lines directly found in the hydrogen spectrum shows that this cannot be accidental but is a real regularity. The vibration frequencies ν of the hydrogen lines (inclusive of the infra-red) may be represented by the formula

$$\text{vibration frequency} = R\left(\frac{1}{n^2} - \frac{1}{m^2}\right)$$

where R is the Rydberg constant $3\cdot291 \times 10^{15}$ and for m are to be inserted successively the integers 3, 4, 5, and so on, while n has the value 2 for the ordinary hydrogen lines and the value 3 for the infrared lines ; in the latter case m begins from 4. All the lines obtained for a given n, when m assumes consecutive whole number values, are grouped together as a series and are called *series lines*.

The attempt was soon made, of course, to apply Balmer's law to other spectra than that of hydrogen. It was found, however, that with them no such simple law existed, no law in which simply the squares of the consecutive integers occurred. But it was seen, nevertheless, that even in these other spectra quite a number of lines were related and that they could be grouped together by making a number m assume all integral values, only m did not occur squared, as with Balmer, but in another function. All such related lines of a spectrum are grouped together as a series ; such series have been found in many other spectra, and it has been shown that all lines which belong to one and the same series are also of similar appearance. In the spectra of the majority of substances several different series are distinguished. The lines of one series are without exception sharply bounded (consisting frequently of two or three closely adjacent lines each), those of another series are diffuse towards the red end, those of a third series diffuse towards the violet end.

A regularity has thus been discovered in the spectra of hydrogen and of many other elements, but a regularity of a peculiar kind. The vibration frequencies of the different lines depend on the successive whole numbers but they are not, like the harmonic overtones of acoustics, proportional to these whole numbers but contain them in the denominator, as the square in the simplest case but otherwise as a more complicated function. Now how can we imagine a vibrating mechanism, i.e. a mechanism with periodic movements, in which so remarkable and peculiar a connection exists between the possible vibrations ? If we could devise such a mechanism it would represent the internal movements of an atom. Many attempts have been made in this direction by mathematicians and physicists but practically in vain.

An extension of Rutherford's nuclear atomic theory to a theory of the whole atom required, of necessity, that it should give a representation of what the movements were which occasioned Balmer's law in the simplest case and the law of series spectra in other cases. The Danish scientist Niels Bohr succeeded, in 1913, in devising such an extension of Rutherford's atom, postulating movement of the electrons round the nucleus in such a manner as completely to meet the above-mentioned requirements in some cases. In spite of many difficulties and obscurities which it still contains, this theory shows so many unexpectedly brilliant concordances with fact that it may safely be assumed to contain a large measure of truth.

In order to understand this theory we must first

discuss a fact which made its appearance in physics as early as 1900, but is still not understood. It is known that in all physical and chemical processes the law first stated by Helmholtz is always obeyed, namely, that energy is neither created nor destroyed in them, but only transmuted from one form into another. By energy is understood everything which results from work or which can be transformed into work. Energy may assume very different forms ; it may be kinetic energy (*vis viva*), potential energy, thermal, electrical, chemical, or luminous energy, and every physical and chemical occurrence is always associated with a transformation of energy from one form into another. We here suppose energy to be something which can vary continuously, i.e. which can increase or decrease by minimal amounts. In all transformations of energy with which we generally have to deal, such a continuous transformation of energy takes place. When, e.g. running water drives a turbine and the turbine sets a dynamo in motion, which delivers an electric current, which supplies the incandescent lamps of a town, the kinetic energy of the water is changed into the energy of motion of the turbine and of the dynamo, and this is converted into the energy of an electric current, and the latter finally into thermal and luminous energy. Ever so small an amount of running water, ever so little work done by it, is eventually changed into a corresponding amount of energy of heat and light. We express this by saying that energy can vary continuously, there are no " jumps " in the energy of the running water nor of the electric current nor of the light.

This idea and this expression were only correct, however, so long as we had to deal with the gross phenomena which matter generally exhibits to us. When, in the last decade, physics has so far advanced that it was able to study the behaviour of separate atoms and electrons, it was found that this idea was no longer correct. The energy which is emitted by individual atoms in the form of vibrations does not behave continuously or uniformly, but varies actually in " jumps " or as we now call it by *quanta*. An atom can emit, as vibrational or radiant energy, a certain minimum amount or *quantum of energy* or two, three, or more such quanta, but it cannot emit $\frac{3}{4}$ or $1\frac{1}{2}$ or $2\frac{1}{4}$ quanta of energy nor any amount at all which is intermediate between one and two such quanta. This conception was first introduced into physics through the investigation of the radiation of a black body, and it then showed itself justified in other cases as well, where vibrations or periodic movements of atoms were in question, particularly in the theory of specific heat. These quanta of energy are not, however, equal in amount for every kind of vibration ; on the contrary they are the greater the higher the frequency. The ratio between the amount of one of these quanta of energy and the frequency of vibration of an atom is a constant ; it plays an especially important rôle in the investigation of atomic processes and is called *Planck's constant* or the " *Wirkungsquantum.*" [1] In absolute measure (cm., gm., sec.) this " Wirkungs-quantum " has the value $6·55 \times 10^{-27}$. Hence it follows that the quantum of energy (h times

---

[1] This constant is usually referred to and written as " *h.*"—TR.

frequency) for periodic movements of different kinds
has values which are given in the following table :—

Nature of Vibration.	Red.	Violet.	Extreme Ultra-Violet.	X-Rays.
Wave-length	7600×10^{-8}	3800×10^{-8}	1000×10^{-8}	1×01^{-8} cms.
Frequency	0.04×10^{16}	0.08×10^{16}	0.3×10^{16}	300×10^{16}
Quantum of Energy	0.26×10^{-11}	0.52×10^{-11}	1.95×10^{-11}	1950×10^{-11}erg.

We have at present no definite idea of the meaning
of Planck's quantum or of the special reason why
atoms emit (and perhaps also take up) vibrational
energy only in separate quanta.   The occurrence of
these quanta in processes of atomic vibration is a
fact deduced from experimental observations, and
must always be taken into account in such processes,
but a clear and detailed explanation of it is still
lacking.

Now these quanta of energy play an important
part in Bohr's atomic theory ; the theory is a further
extension of Rutherford's nucleus theory with the
aid of the " Wirkungsquantum." On Bohr's theory
an atom corresponds with a regular solar system.
Every atom has a positively electric nucleus of very
small dimensions containing as many free positive
elementary charges as are required by the atomic
number of the element.   Around this positive nucleus the negative electrons revolve in circles (or,
more generally, in ellipses), just as the planets revolve
round the sun. The force which compels each electron to remain in its orbit, the centripetal force, is
the simple Coulomb force of attraction between the
positive charge of the nucleus and the negative charge
of the electron.   In the presence of other negative

electrons than that under consideration, this attraction is diminished by the repulsive forces of the remaining negative electrons. Under the influence of this attraction and of an original velocity which has been imparted to it, say, by the impact of another electron, the electron continues in its circular or elliptical orbit, just as the moon does in its circle round the earth or the earth in an ellipse round the sun. The centrifugal force of the moving electron is always equal to the centripetal force exerted on it by the nucleus (and by other electrons if present).

But now the planets revolving round the sun are at very different distances from it and have very different velocities in their orbits, and their distances may be of any magnitude without restriction ; this is not the case with the electrons revolving about the nucleus, because of the " Wirkungsquantum." On the contrary, they can only travel in orbits round the nucleus at certain particular distances defined by whole numbers, and their velocities of revolution must have particular values.

Let us take the simplest case, that of the hydrogen atom, in which there is only one positive nucleus with a unit charge and only one negative electron with the same (negative) charge. If the electron is in motion with unknown velocity in a circle of unknown radius, there are two equations connecting the two unknowns, the radius and the velocity. Firstly, the Coulomb force of attraction must be equal to the centrifugal force, and secondly, the kinetic energy of the electron must be equal to an integral multiple of Planck's constant " h," divided by the time for two revolutions. From the former relation we find that the square of the velocity which an electron has on a possible path of a certain radius is equal to the square of the elementary charge existing both on the nucleus and on the electron, divided by the mass of the electron and by the radius of the circle in which it moves. From the second relation it follows that the velocity of the electron in this circle is equal to this integral multiple of Planck's constant, divided by the mass of the electron and by the radius of the circle (multiplied by 2π). From these two relations both the radius of the orbit and the velocity may be calculated, provided the two charges (of the nucleus and of the electron), the mass of the electron and Planck's constant are given. According to the particular integral multiple of h, we say that an electron moves on a one-quantum, two-quantum, three-quantum orbit, and so on.

Now, as to the radius of the first or one-quantum orbit, it is found from the given values that
Radius of first orbit

$$= \frac{\text{square of } h}{\text{mass of electron} \times \text{square of elementary charge}} \times \frac{1}{4\pi^2}$$

Neglecting elliptic orbits and considering only circular orbits for the sake of simplicity, the radii of the successive possible circular orbits must, on account of the quantum, be related to one another as the squares of the successive whole numbers, i.e. as $1 : 4 : 9 : 16 : 25 : 36$ and so on. If we introduce the values for h, for the mass of the electron, and for the square of the elementary charge from pages 47 and 33, we find that in the case of hydrogen the innermost circle, the one-quantum orbit, has a radius 0.55×10^{-8} cms., the second 2.20×10^{-8}

cms., the third 4.95×10^{-8} cms., the fourth 8.80×10^{-8} cms. and so on; the first two of these values are comparable with the radius of the sphere of influence of the hydrogen atom (2.5×10^{-8} cms. according to p. 15).

The velocity of the electron in the first orbit is similarly given by the above equations :
velocity in first orbit

$$= \frac{\text{square of elementary charge}}{h} \times 2\pi$$

This velocity amounts to 2.172×10^8 cm./sec., i.e. about $1/138$ of the velocity of light. In the succeeding orbits the velocity is less ; the farther the orbit is from the nucleus the slower the motion of the electron, the velocities in the orbits being related to one another as $1 : \frac{1}{2} : \frac{1}{3} : \frac{1}{4} : \frac{1}{5}$, etc.

In order completely to separate an electron from the nucleus and to remove it to an infinite distance from the nucleus, work must be expended on, or supplied to, the system, since nucleus and electron attract one another. The work required is the greater, the nearer the electron is to the nucleus, i.e. the smaller the radius of the orbit : this work is always equal to the kinetic energy of the electron, i.e. to the product of half the mass by the square of the velocity of the electron. We shall designate as the *work of separation* of the electron the work which is necessary to remove the electron from the first orbit to infinity. From the above-given value for the velocity in the first orbit we therefore obtain at once work of separation

$$= \frac{\text{mass of the electron} \times (\text{elementary charge})^4}{\text{square of } h} \times 2\pi^2$$

If we insert the values, mass of electron $= 9\cdot0 \times 10^{-28}$ (p. 47), elementary charge $= 4\cdot74 \times 10^{-10}$ (p. 33), $h = 6\cdot55 \times 10^{-27}$ (p. 129), the work of separation for the hydrogen electron comes out to $20\cdot90 \times 10^{-12}$ ergs.

Since the velocities in the successive orbits decrease in the ratios $1 : \frac{1}{2} : \frac{1}{3} : \frac{1}{4}$ and so on, the work done in removing an electron from one of these orbits to infinity is smaller than the work of separation in the ratios

$$\frac{1}{1^2} : \frac{1}{2^2} : \frac{1}{3^2} : \frac{1}{4^2} \text{ and so on.}$$

Suppose now that we have a cubic centimetre of hydrogen; in it there are trillions of nuclei and trillions of electrons. We will suppose that both the nuclei and the electrons are in irregular thermal motion. Then, once in a while, some electron will, in virtue of an impact which it receives, enter on one of the possible orbits close to a nucleus and will rotate in this orbit. Another impact will expel it out of this orbit and make it go to a smaller or greater distance from the nucleus, where it will then describe an orbit of smaller or greater radius. If in so doing it goes from a more remote to a nearer orbit, it must give up energy. If, for example, an electron passes from the third orbit to the second it gives up, in so doing, a quantity of energy which is equal to

$$\text{work of separation} \times \left(\frac{1}{2^2} - \frac{1}{3^2}\right).$$

And if it passes from the fourth to the second orbit the energy given up is

$$\text{work of separation} \times \left(\frac{1}{2^2} - \frac{1}{4^2}\right).$$

Now on the quantum theory this amount of energy given up can only be a quantum of energy, and each quantum of energy given up corresponds with a frequency of vibration determined by its amount: this frequency determines the colour of the light radiated in this process. Since, according to page 129, the vibration frequency is equal to the quantum of energy divided by h, it is obvious that if an electron passes from the third orbit to the second the frequency of vibration of the radiation must be equal to

$$\frac{\text{work of separation}}{h} \times \left(\frac{1}{2^2} - \frac{1}{3^2}\right).$$

and if it is driven from the fourth orbit into the second the frequency is

$$\frac{\text{work of separation}}{h} \times \left(\frac{1}{2^2} - \frac{1}{4^2}\right).$$

Thus precisely the same factors occur as those which appear in Balmer's formula for hydrogen. Not only is this so for Balmer's factors, but the fraction, work of separation divided by h, is found on calculation to be closely equal to Rydberg's constant R. For from the work of separation $20 \cdot 9 \times 10^{-12}$ erg. (p. 134) and the " Wirkungsquantum " $6 \cdot 55 \times 10^{-27}$ we have

$$\frac{\text{work of separation}}{h} = \frac{20 \cdot 90}{6 \cdot 55} \times 10^{15} = 3 \cdot 191 \times 10^{15}.$$

This agrees extremely well with the experimentally determined value of Rydberg's constant $R = 3 \cdot 290 \times 10^{15}$. When, therefore, an electron flies from the mth orbit to the second, a vibration results whose frequency is

$$\nu = R\left(\frac{1}{2^2} - \frac{1}{m^2}\right).$$

which is exactly the frequency corresponding with Balmer's formula for the hydrogen series. This is obviously a very striking achievement which gives supplementary confirmation of the general idea.

Rydberg's constant is thus equal to

$$\frac{\text{work of separation}}{h}$$

and, from the value given on page 133 for the work of separation, it becomes

$$\frac{\text{mass of an electron} \times (\text{elementary charge})^4}{h^3} \times 2\pi^2.$$

Among the trillions of nuclei and electrons present in a cubic centimetre of hydrogen all possible modes of movement will very often occur. Electrons will be shot many thousands of times from the third orbit to the second, and many thousand more from the fourth or fifth orbit to the second. The sum total of the simultaneous and successive occurrences of such motion is what we see as the spectral lines radiated by hydrogen.

The question must at once be asked, why are these electrons pushed only into the second orbit, why is this orbit preferred? Why are not the spectral lines seen which arise when an electron is thrown from the third or fourth orbit to the first, or from the fourth or fifth to the third orbit? The answer is simply that such cases must, of course, also occur just as often as a transfer to the second orbit, but the spectrum lines resulting would not fall in the visible range, which only extends from 7600 to 3800 Å.U.

As a matter of fact, the differences which would

occur for the transfer from a more remote orbit to the first are

$$\frac{1}{1^2} - \frac{1}{2^2}, \quad \frac{1}{1^2} - \frac{1}{3^2}, \quad \frac{1}{1^2} - \frac{1}{4^2}, \quad \frac{1}{1^2} - \frac{1}{5^2}.$$

They have therefore the values

$$\frac{3}{4}, \quad \frac{8}{9}, \quad \frac{15}{16}, \quad \frac{24}{25}.$$

The Rydberg constant having the value $3 \cdot 290 \times 10^{15}$, the vibration frequencies pertaining to them are thus

$$2 \cdot 4675 \times 10^{15}, \ 2 \cdot 9244 \times 10^{15}, \ 3 \cdot 0844 \times 10^{15}, \ 3 \cdot 1584 \times 10^{15}$$

and the corresponding wave-lengths

1216 Å.U., 1026 Å.U., 973 Å.U., 950 Å.U.

These waves thus fall in the most extreme ultra-violet, which is very difficult of observation simply because these short-waved rays are very strongly absorbed even by the air. They have been observed by Lyman.

Again the differences which would occur for transfer from a more remote orbit to the third, namely,

$$\frac{1}{3^2} - \frac{1}{4^2}, \quad \frac{1}{3^2} - \frac{1}{5^2}, \quad \frac{1}{3^2} - \frac{1}{6^2}, \quad \frac{1}{3^2} - \frac{1}{7^2}.$$

have the values

$$\frac{7}{144}, \quad \frac{16}{225}, \quad \frac{1}{12}, \quad \frac{40}{441}.$$

The corresponding frequencies are thus

$$0 \cdot 15998 \times 10^{15}, \ 0 \cdot 23403 \times 10^{15}, \ 0 \cdot 2742 \times 10^{15},$$
$$0 \cdot 2984 \times 10^{15},$$

and the corresponding wave-lengths

18,752 Å.U., 12,819 Å.U., 10,941 Å.U., 10,053 Å.U.,

so that they fall in the infra-red and are only recognizable by special instruments. The first two lines of this series have actually been observed, as mentioned above on page 125.

The difference $\frac{1}{4^2} - \frac{1}{5^2}$ and those corresponding with it, for the transfer into the fourth orbit, give lines which lie much farther still in the infra-red ; the first would have a wave-length of 40,527 Å.U. Thus only those lines are in the visible spectrum which result from a transfer from a more remote orbit to the second. And such transfers are to be observed, according to Balmer's formula and the verification of it, even from the thirty-first orbit. But here it is to be noted that, in general, the movement of electrons at greater distances from the nucleus, say in the tenth or fifteenth orbit, can only occur under peculiar conditions. These distances are so great as to be more than the mean distance between two nuclei at the density of gas usually occurring in Geissler tubes. The fifteenth orbit has, in fact, a diameter of 2.5×10^{-6} cms. according to page 132. At atmospheric pressure the mean distance between two atoms, according to page 16, is some 3.3×10^{-7} cms. ; on rarefaction to the pressure of one or two millimetres which prevails in spectrum tubes, the distance would be some $\sqrt[3]{760}$ times as great, i.e. 2.7×10^{-6} cms., which is of the same magnitude as that of the fifteenth or sixteenth orbit. An electron at a greater distance from a nucleus would consequently describe its orbit, not about this, but about the next nucleus. Hence it may be inferred that the ultra-violet spectrum of hydrogen in a Geissler tube can only be followed as far as the twelfth line of the Balmer series, and this is actually the case. The other lines up to No. 29 have, as a matter of fact, been observed, not in Geissler tubes, but only in the spectrum of the *nebulæ*, in which the density of the gas must be supposed to be very low, the intensity of the lines being nevertheless great in virtue of the great area of the nebulæ. This corresponds exactly with what would be expected on Bohr's conception.

At atmospheric pressure a molecule of hydrogen has a space to itself equal to a cube of which half the side is 16.5×10^{-8} cms. Into this space only orbits 1 to 5 would go, the last having a radius of 13.7×10^{-8} cms. The fifth orbit corresponds with the hydrogen line $H\gamma$. It would therefore be expected that at atmospheric pressure hydrogen would show only the lines $H\alpha$, $H\beta$, and $H\gamma$.

In order to obtain an idea of the orders of magnitude in a hydrogen atom so constituted, let us again suppose the hydrogen atom of radius 1×10^{-8} cms. magnified until it occupies the volume of the whole earth, i.e. till it has a radius of 6350 kms. Then the nucleus has a radius of 6 cms. corresponding thus with the size of a child's ball. The electron has a radius of 120 m. corresponding with the size of a church or barracks. This barracks then rotates round the child's ball and the smallest distance of the electron from the nucleus (in the first orbit), 0.55×10^{-8} cms., corresponds therefore with about half the radius of the earth, a distance of 3500 kms. The other orbits have radii twice, four and a half times, etc., that of the earth. We see that only very little of the whole space in the atom is actually occupied by (electrical) matter, the greater part of it being void.

On the other hand we observe that an atom has really no definite volume, but that its volume is very different according to the rarefaction of the gas. In the very tenuous nebulæ an atom has a radius which is about 1000 times as great as that of the innermost ring, for there the nucleus continues to act on the electron even up to this distance.

The most stable condition is when the electron is revolving in the first orbit, i.e. in that nearest to the nucleus. In this case the radius of the atom would be only about one quarter of the radius of the sphere of influence calculated for the molecule on the gas theory.

No other atom, of course, is so simply constructed as that of hydrogen. The next simplest is that of helium in which the nucleus has two positive charges and which must therefore contain also two outer electrons. For the arrangement of the electrons there are at once several possibilities. Each of them might revolve in a separate orbit or they might both move in the same orbit at a distance, for reasons of symmetry, of a semi-circle apart. These possibilities we shall discuss in the next chapter. The vibration frequencies which result for the motion of one of these electrons from an outer to an inner orbit are subject to the same rule as Balmer's hydrogen frequencies except that, since the nucleus has here two unit charges, it is not the integers themselves but their halves which appear in the denominator as the square. A series of helium lines has, in fact, been observed which corresponds with this law, though the principal lines mentioned on page 120 do not belong to it. Rydberg's constant appears again for helium. But no one has yet been able to make a complete calculation of all the possible vibrations for the helium atom, as has been done for hydrogen ; in particular, as we have just remarked, the occurrence of the principal lines in the visible spectrum has not yet been explained.

Series have been found purely empirically for all other spectra, especially for the alkali and alkaline earth metals ; these series are of different kinds, but all are characterized by the variation of a quantity by whole numbers. In all these series the Rydberg constant, or at any rate a close approximation to it, has always reappeared on calculating out the experimental data. It is thus a fundamental constant which occurs in all series vibrations. On Bohr's theory, too, this constant should recur in many cases. For even when the nucleus is made up of many units with the same number of electrons circulating round it, the behaviour at a great distance from the nucleus must still be just the same as for the attraction of the hydrogen electron by its nucleus, for at this distance the forces on an electron again depend essentially on one elementary charge (the nuclear charge *minus* the charges of all the other electrons) : hence the Rydberg constant must re-appear for this case.

Incidentally, it is found, in constructing the empirical formulæ for the series lines in such cases, that though the constant which appears is always very nearly equal to the Rydberg constant for hydrogen, yet there are certainly small deviations from this value. But Bohr's theory at once provides an explanation for this fact as well. If an electron is attracted by a nucleus and moves in a circle, the centre of this circle lies, strictly speaking, not at the nucleus (supposed to be a point) but, since the nucleus is also being attracted, at the centre of gravity of the masses of the electron and the nucleus. According to the ratio of the two masses, which is always very small, the constant of the series lines for such atoms must therefore show small deviations from Rydberg's number for hydrogen.

A detailed application of Bohr's atomic theory to these complicated cases encounters considerable difficulties. It has to allow for the possibility that, at the small distances from the nucleus which are here in question, Coulomb's law of attraction may not hold exactly as if the nucleus were a point, but that on the contrary the shape of the nucleus may then be a factor, e.g. if the nucleus were shaped like a flat disc instead of a sphere, quite different laws would hold for attraction at small ranges. There is here a fruitful but stubborn field available for an extension of the theory, which is not limited to providing numerical results but endeavours to fathom the obvious facts of physics.

Sommerfeld has succeeded, by an amplification of Bohr's theory, in explaining what is called the *fine structure*, i.e. the numerous fine lines which are found at high dispersion in the spectra of hydrogen and of helium.

If the transfer of an electron from a more remote to a nearer orbit is connected with the emission of radiation, the absorption of radiation must, conversely, be connected with the transfer of an electron from an inner to an outer orbit.

The transfer of an electron from the innermost orbit to infinity requires an expenditure of work or an addition of energy which we have denoted, on page 133, as work of separation and have calculated for the hydrogen atom. Now such a removal or separation of an electron from the nucleus is precisely what is designated as *ionization* of the atom. From the neutral hydrogen atom, nucleus + negative electron, there is thus formed a hydrogen ion, which is just the nucleus. The electron is completely separated from the nucleus. Such ionization of atoms is always, in fact, produced when electrons or ions in rapid motion are caused to collide with atoms, the process being known as *ionization by collision*, or when X-rays or gamma-rays are allowed to strike the atom, and there are also other methods. In ionization by collision with moving electrons, the colliding electron must, in order that it shall be able to ionize, have sufficient kinetic energy to supply the work of separation to the atom which it hits. In actual cases the colliding electron always obtains this kinetic energy by being more and more accelerated under the influence of a difference of potential, until it has attained the requisite velocity. The potential difference, to the action of which the colliding electron is subjected, supplies to it such an amount of work that the kinetic energy resulting is, in its turn, sufficient to provide the work of separation for the electron which it hits. This potential difference is designated as the *ionization potential*. But this means that the work of the ionization potential on the colliding electron must be equal to the work of separation of the electron hit. Now when an electric potential (measured in volts) moves a number of coulombs, the work is equal to the product of the volts and the coulombs, and this work is then expressed in 10^7 ergs. The number of coulombs moved is here the elementary charge of the colliding electron, which is 1.58×10^{-19} coulombs according to page 33. Hence it follows that

ionization potential $\times\ 1.58 \times 10^{-19}$ coulombs $\times\ 10^7$
$= $ work of separation (in ergs).

For the ionization potential of the hydrogen atom we thus have, according to page 134,

$$\text{ionization potential} = \frac{20.91 \times 10^{-12}}{1.58 \times 10^{-12}} = 13.2 \text{ volts.}$$

This number is rather greater than that found experimentally for the ionization potential of hydrogen, namely, 11 volts. The reason for the difference is probably that it is not certain, in the experimental measurement, whether the separated ion has really been completely removed from the nucleus or only taken to a greater distance from it. In the latter case the ionization potential must be correspondingly less than 13.2 volts. In any event the Bohr atom again shows close agreement with facts in this respect, and the same holds not only for the hydrogen atom, but also for other atoms, as has been demonstrated by the measurements of Franck and Hertz and of others.

When Bohr's theory is extended from the simplest cases of hydrogen and helium, with one or two electrons, to the more complicated cases in which a large number of electrons are in motion about the nucleus, a plausible assumption can be made which is supported by several considerations ; this is that a number of electrons revolve about the nucleus in one and the same ring, at equal distances from one another. One ring only or several successive rings may be thus studded with electrons. It is not in all cases that such a system will have stability ; only those arrangements are possible in which the system can continue stable.

Radiation may then be produced either by the whole of an external ring intruding itself into an inner one and coalescing with it—which only leads to stable arrangements if both the rings have the same number of electrons—or by a single electron jumping from an outer to an inner ring.

The determination of all the possible vibrations which can be executed by so complicated an atom is, of course, a problem of much difficulty and has not hitherto been solved in any case. Nevertheless it may readily be shown that the very rapid vibrations of such an atom, the high-frequency vibrations, must follow Moseley's law (p. 108).

The determining factor in the calculation of frequencies is always the difference of energy between the first and the second state (the more remote and the nearer ring). The energy of a moving electron depends upon the square of its charge and upon the square of the charge which acts upon the moving electron. The square root of the frequency must thus be proportional to the charge of the electron (which is constant) and to the charge of the system acting upon it. If then there is, in a

10

complicated atom, a ring of electrons about the central nucleus which has, say, an N-fold charge, the charge which acts upon an electron in this ring is not N times the elementary charge, but is less, simply because the other electrons of the ring exert upon it a repulsion instead of an attraction. The ring, therefore, acts as if there were in the nucleus instead of its N-fold charge a smaller one $(N - a)$, where the number a depends in a simple manner on the number of electrons in the ring, e.g. with two electrons in the ring, $a = 0.25$, with three electrons $a = 0.57735$, with four electrons $a = 0.9571$. But it follows from this that, when an electron jumps from one ring to another, the square root of the frequency is proportional to $N - a$, where N is the atomic number of the particular atom in the periodic system ; for the charge on the nucleus is proportional to this atomic number.

But in this statement we have derived the law found by Moseley for all X-ray spectra, that the square root of the frequency is a linear function of the ordinal number of the element. This remarkable law is therefore completely explained by Bohr's model of the atom.

Nay more, it is possible, at least to some extent, to determine, by comparison of the observations on the frequencies of the K-lines (in the first place of the K_{a_1}-line), which ring is concerned in them. Bohr h ound in this manner that the K_{a_1}-line is produced by an electron moving from the second ring into the first, the first ring then consisting, in its stable condition, of four electrons. Similarly the L_{a_1}-line results from the motion of an electron from the third to the second ring. The occurrence of the other lines $K\alpha_2$, $K\beta$, etc., is explained (according to Sommerfeld) by similar movements.

These particular deductions are, however, by no means safe. According to Debye, the occurrence of the $K\alpha_1$-line may also be explained by supposing that the innermost ring in all atoms (from sodium upward) is, in the stable condition, a ring with three electrons. When, for any reason, one of these three electrons is transferred to the two-quantum ring and then falls back again, so that the stable arrangement is reproduced, the $K\alpha_1$-line results, and correspondingly the $K\beta_1$-line results when the electron falls from the third ring back into the innermost. The inside ring of all the atoms from sodium onwards (the K-lines have not been observed below this) would accordingly always be formed, in the stable condition, of three electrons. Kroo, in a continuation of this investigation, has made it probable that the two-quantum ring (again from sodium onwards) always consists, in the stable state, of eight electrons.

It is evident, from what has been said, that Bohr's model of the atom embraces, in a very simple manner, a number of important and remarkable experimental facts. The explanation of the series lines, in particular the complete explanation of Balmer's series for hydrogen, the derivation of Rydberg's constant, and the deduction of Moseley's law for X-ray spectra, are very striking achievements of this theory of the atom. There are, however, many experimental facts which still await more detailed explanation on this theory. In particular, there are a number of regularities which have been found in the series spectra of the alkali and alkaline earth metals, and others also which have been observed in the action of a magnetic field on the series lines (known as the Zeeman effect) which still require accurate explanation by this model. On the other hand, the action of an electric field on the series lines (known as the Stark effect) can be quite completely elucidated by Bohr's model of the atom. For the rest, it should not be overlooked that there are several arbitrary assumptions made by Bohr, which only served to derive Balmer's series. According to our present knowledge of electricity an electron cannot, in general, continue to rotate in a circle about a nucleus ; for it is giving up energy to the ether all the time and must, in consequence, gradually fall into the nucleus. But Bohr's theory supposes that, in the orbits described, no radiation of energy into the ether takes place. For this assumption, as for a number of others, there is no *a priori* foundation ; it is warranted only by its splendid result, the explanation of the series lines, and by the fact, which is incontrovertible, that in the domain of the atom, at minimal distances, the laws of mechanics and electro-dynamics which we have found elsewhere are not of general validity.

LECTURE VI

FURTHER INVESTIGATIONS OF THE STRUCTURE OF NUCLEI, ATOMS, IONS, AND MOLECULES. THE DECOMPOSITION OF NUCLEI

CHEMISTS have often been likened to architects, for both have as the object of their work the disposition and assembling of structures, the one on the large scale and the other in the domain of the atom, and both need for their task artistic intuition in addition to scientific grounding in order to recognize the right and the true amongst a host of possibilities. This same artistic intuition will be necessary in order to investigate for all cases the actual construction of atoms from nuclei and electrons. For there are only very few clues to the solution of the problem in cases where any considerable number of electrons is concerned. One of these clues is, that stable configurations require a larger expenditure of energy to cause their destruction than the less stable. If, then, the amount of energy is calculated which is necessary for the dissolution of one configuration and this amount is compared with that for another, it may at once be seen which of the two configurations is the more probable ; it will be that for which the energy required is the greater. This criterion is not, however, sufficient in all cases, it may in fact be misleading in many, for even a relative maximum of energy may produce stable conditions while larger values of the requisite energy may, though they are not maximal, yet lead to labile states.

A second clue is that provided by chemistry, which shows that the valencies of atoms differ and that atoms have in general 0, 1, 2, 3 or 4 valencies. On our assumptions this means that there must be in these cases 0, 1, 2, 3 or 4 electrons more weakly or more loosely bound to the nucleus than the others. It is by means of these more loosely bound electrons that an atom fastens itself to another. Now chemistry teaches us, further, that the valency of an atom may be variable, that under certain circumstances one and the same atom may be divalent, under other circumstances trivalent, and it may have even more than four valencies, five, six, seven or eight. A complete theory of the structure of any particular atom will have to make clear, from the arrangement of the electrons, the reasons for this polyvalency.

There have, it is true, been many attempts made to this end, but hitherto they have been very imperfect and mutually contradictory. It is a problem for the future to gain truer knowledge here. The chemistry of the composition of atoms, the superchemistry, has to deal with the same enormous number of possibilities as has ordinary chemistry, the chemistry of molecular composition. Just as the latter succeeded only gradually, by advancing from the simple to the more complex, in making clear the constitution even of very complicated molecules, so in the study of the constitution of atoms there can only be gradual success in establishing their actual structure.

But here there is yet another and more fundamental problem. In the chemistry of the composition of atoms the aim is to find the arrangement of the external electrons round a given atomic nucleus. But in most atoms the nucleus itself is composite. We know that in radio-active substances it consists of helium nuclei and negative electrons, and we must suppose that most other atoms are similarly compounded. Science has therefore the further problem of determining the composition of the nuclei. Since the latter problem can be attacked in the first place chiefly by physical methods, as has been the case with radio-active substances, we may speak of a physics of the nucleus. Such a nuclear physics would lead us, if we could make any sort of start on it, to the inmost recesses of the material world. It would lead direct to an analysis, and perhaps also to a synthesis, of atoms and so enable the dream and hope of the old alchemists to be realized in a newer but scientifically more exact form.

In the last few years nuclear physics has been advanced in the most remarkable and brilliant manner by Rutherford's ingenious methods of experimentation, and it has given results which no physicist or chemist would have suspected a short while ago.

It is radio-activity, the field in which Rutherford has been the leading worker and pioneer, which is directly concerned with atomic nuclei and their disintegration : and from it have come to light the most recent sensational results.

The a-particles, which are shot out from radium or from radium-A, B or C, and which consist, as we saw on page 63, of helium nuclei, have a range, in air, of 3-7 cms. according to their velocity. The α-particles are easily detected by exposing to them a screen of zinc sulphide : each particle then produces a momentary flash on the screen at the place where it strikes it. The screen is therefore seen to flash first in one place then in another ; there is what we call a *scintillation* of the screen, and this only takes place when the screen is at a distance of 3-7 cms. from the source of the α-particles. Closer observation, however, was first made by Marsden and was then followed up by Rutherford : this showed that occasional weak scintillation of the screen occurred at much greater distances, even up to 30 cms. These points of scintillation gave the impression that they were not due to α-particles, i.e. to doubly charged helium nuclei, but to something quite different ; it seemed possible that they might be due to hydrogen nuclei.

This suggestion of Rutherford's was shown to be justified by a very difficult investigation brilliantly carried out by him. He subjected these flying particles of unknown nature both to magnetic deflection in a magnetic field and to electrostatic deflection in the field of a condenser. We pointed out on pages 42 and 43, that it was possible, from a measurement of the deflections so produced, to determine on the one hand the velocity, and on the other hand the specific charge (p. 32) of the unknown particles. In this manner Rutherford found from his measurements that the specific charge of these unknown corpuscles is about 100,000 coulombs/gramme. Remembering that hydrogen in electrolysis has a charge of 96,494 coulombs/gramme (p. 33), we see that, within the limits of accuracy of the very difficult observations, we have here the same number; it may safely be concluded that the particles which are here in question are hydrogen atoms with a single positive charge, i.e. *hydrogen nuclei*. Their velocity was found in the same experiments to be 1·6 times as high as that of the helium nuclei, and their range, which depends on the velocity, was, as has been mentioned, some four times as great as that of the α-particles.

The first idea was, naturally, that these hydrogen nuclei were constituents of the radium (radium-C was used) just as the helium nuclei are, and were set free from it. But this idea could not be verified. It was also possible that they might have been set free from moisture still present in the air in spite of all its drying, or else from other substances containing hydrogen. But it was established by counting that there is only one such hydrogen nucleus produced for every 100,000 α-particles of 7 cms. range, and calculation showed that a hydrogen nucleus is not set free from an atom of hydrogen until an α-particle approaches within some 3×10^{-13} cms. of it, i.e. gets quite close to it.

Now Rutherford examined also the phenomena which occur when α-particles enter into pure oxygen or pure nitrogen. In the latter case a very unexpected phenomenon was observed. When α-particles entered pure dry nitrogen, very many more nuclei of long range, i.e. presumably hydrogen nuclei, resulted than could have been expected. In pure oxygen, or in pure carbon dioxide, hydrogen nuclei were also produced, it is true—from the moisture always present—but they were in the limited numbers above mentioned. But in nitrogen, under certain circumstances, nearly double as many such nuclei occurred. This could not have had anything to do with the moisture content, for it was immaterial whether the nitrogen was dried as completely as possible or was purposely used very moist. The only explanation is one which is of surpassing interest, namely, that the nitrogen nuclei actually contained these long-range nuclei and that, owing to the impact of the α-particle, one of these nitrogen nuclei is disintegrated so that the long-range nucleus comes out of it free.

Whether they are really hydrogen nuclei that here occur could not be stated without new experiments. The long range does not directly show it. For singly charged particles with masses of 2, 3 or 4 would also obtain such velocities, from the impact of an α-particle, as to have about four times the range of the latter.

The question could only be solved by measuring the deflection of the particles in a magnetic field. In spite of the enormous difficulties in the way of such a measurement it was carried out by Rutherford, by making a direct comparison between the deflection of these particles in the magnetic field and that of the α-particles in the same field and under the same conditions. Now the extent of the deflection of a particle in the magnetic field depends on the ratio $\frac{\text{specific charge}}{\text{velocity}}$. The specific charge of an α-particle of charge 2 and mass 4 is equal, in certain units, to $\frac{1}{2}$. For singly charged particles of masses

1, 2, 3, and 4 it is, in the same units, 1, $\frac{1}{2}$, $\frac{1}{3}$, and $\frac{1}{4}$. Assuming the conservation of momentum and of energy, the velocities of the particles with masses 1, 2, 3, and 4 when struck by an α-particle are 1·6, 1·33, 1·14, and 1 times as great as that of an α-particle. We therefore have for the masses

$$\frac{\text{specific charge}}{\text{velocity}} = \frac{1}{1\cdot6},\ \frac{1}{2\cdot7},\ \frac{1}{3\cdot4},\ \frac{1}{4},\ \left|\ \frac{1}{2}\right.,$$

with column headings 1, 2, 3, 4 | α-particle,

and it is seen that this ratio, and therefore the magnetic deflection of the particles, is in the first case greater than, and in the other three cases less than, that for the α-particles. Now the experiment showed conclusively that the magnetic deflection of the long-range particles from nitrogen is greater than that of α-particles, whence it follows without a doubt that they are singly charged hydrogen nuclei.

Whereas in the heavy atoms of radio-active substances there are helium nuclei present, we have here so light an atom as that of nitrogen also showing itself to be composite, and it is found that hydrogen is a constituent of the atom of nitrogen. Nothing of the sort was found in the case of oxygen. It is to be noted that oxygen with the atomic weight 16 can contain exactly 4 helium nuclei with the atomic weight four ; whereas nitrogen, whose atomic weight is 14, cannot be composed solely of helium nuclei but must contain also nuclei of another kind, and this brilliant work of Rutherford has recognized hydrogen nuclei among them. And further it results from energy considerations that though the close approach of an α-particle to a nitrogen nucleus gives rise to the emission of this constituent, the expenditure of energy in the disruption cannot be covered by the energy of the colliding α-particle ; this elimination results from internal atomic forces, just as in radio-active substances. The only difference is that the process is quite spontaneous in the latter case, but bombardment by the α-particle is here required.

In this exposition Rutherford remarks that it is possible that two hydrogen nuclei with one electron might form the nucleus of another atom, which would then have a mass 2 and a nuclear charge 1, so that it would be an isotope of hydrogen. But there might also be a structure, consisting of a hydrogen nucleus and a negative electron, in which the electron was not, as in the hydrogen atom, at a great distance from the nucleus but directly adjacent to it. Such a structure would be electrically neutral except in its immediate neighbourhood. At the same time it would have a certain linear shape and would orient itself in a definite direction in a magnetic or electric field. The possibility is not remote that what we call ether, which is, as is well known, a very bugbear in physics, might perhaps consist of such a doublet. The linear extension of this structure and its orientation in electric and magnetic fields would, in particular, afford an explanation of the fact that ether propagates electric and magnetic disturbances in the form of transverse vibrations. It would then also be conceivable, according to an idea of Nernst's, that under certain circumstances hydrogen atoms might be generated out of the ether, or, conversely, hydrogen atoms might be absorbed into the ether.

In short, speculation finds here a wide field before it, which must be ploughed with patient toil, but will then perhaps yield splendid fruits of knowledge for the subjugation of nature.

Nuclear physics having supplied us with such unexpected and welcome results, there remains less to do in the other study, above suggested, of the chemistry of atoms or the study of the structure of atoms which consist of nuclei and electrons. This will be much less concerned with discovering totally new facts than with making systematic use of the facts already known and improving the accuracy of observations.

What has been established beyond a doubt is, first, that the total number of external negative electrons in a neutral atom must be equal to the net positive charge of the nucleus. But from experiments on positive rays and on radio-activity we know that atomic ions may also exist, atoms, that is, which are charged positively or negatively with one or more charges. Such atomic ions must have either less or more negative electrons within the sphere of influence of the nucleus than have the corresponding neutral atoms. If there are 1, 2, 3 . . . electrons less than in the neutral atom, we have to do with an atom with 1, 2, 3 . . . positive charges. Conversely, if there are 1, 2, 3 . . . more electrons bound to the nucleus than correspond with the neutral atom, there is produced an atom with 1, 2, 3 . . . negative charges. In this connection it will have to be borne in mind that, in electrolysis, metals and hydrogen appear with positive charges, i.e. are electro-positive, so that there may easily be one or more electrons lacking from them, whereas the electro-negative ions like chlorine, oxygen, and so on, have more electrons than correspond with the charge on their positive nucleus. In any detailed examination of the composition of atoms, this variation in behaviour of the different atoms will have to be borne in mind.

In the first four atoms of the periodic system, hydrogen, helium, lithium, and beryllium, the conditions are much simpler than is the case with atoms of higher atomic number. For the former, the stated considerations are in general sufficient to decide both the constitution of the neutral atom and the possibility of atomic ions.

1. HYDROGEN

In the neutral atom, the positive nucleus is accompanied by an electron. When the electron is in the most stable permanent position, i.e. on the innermost one-quantum orbit, it is at a distance of 0.55×10^{-8} cms. from the nucleus, and describes its orbit 6.5×10^{15} times per second. These numbers can be calculated, as we found above (pp. 132 *et seq.*), since there are two equations connecting the radius of the most stable orbit and the velocity, and from these equations the two quantities can be found when the charge of the nucleus, the charge and mass of an electron, and Planck's constant are known. The neutral hydrogen atom is thus of the form shown in Fig. 34.[1]

[1] In this and the following figures the radii of the orbits are shown to scale. The electrons are represented as large spheres, the nuclei as points : this gives an idea of their relative proportions, at any rate in the case of hydrogen. The dimensions of the electron spheres are, however, much exaggerated in comparison with the radii of the orbits.

If the electron is by any means torn away from its nucleus, we have then simply the nucleus, which is a positive hydrogen ion.   This nucleus occurs in the electrolysis of water (dilute acids) and, as already mentioned, in a number of radio-active phenomena.

There   may   also,   however,   be   two   negative

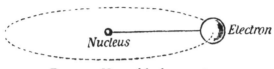

FIG. 34.—Neutral hydrogen atom.

electrons rotating about the nucleus at a rather greater distance.   The energy requisite for the destruction of this system is somewhat greater than for the atom in the neutral state.   We thus have a negatively charged hydrogen atom, such as has been found among positive rays.   The distance of the two electrons from the nucleus is 1·33 times as

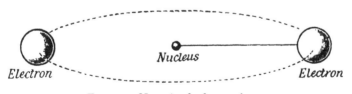

FIG. 35.—Negative hydrogen ion.

great as in the neutral atom ; the number of revolutions per second is, however, only 0·563 times as many.   Fig. 35 represents this negative hydrogen ion.   The hydrogen atom cannot take up more than one negative charge and nothing of that kind has been observed.

2. HELIUM

The nucleus with its double elementary charge forms the helium ion with two positive charges, which is the α-particle of radio-activity. About this nucleus there may be one electron rotating, in which case we have a helium ion with a single

FIG. 36.—Helium ion with one positive charge.

positive charge. The distance of the electron in this case is only half of that in the hydrogen atom as Fig. 36 shows. On the other hand the electron rotates around the nucleus about four times as fast as does that of the hydrogen atom.

The neutral helium atom has, according to Bohr (Fig. 37), two electrons revolving round the nucleus at a rather greater distance (1·15 times that in the

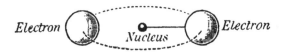

FIG. 37.—Neutral helium atom.

positive helium ion) and with rather smaller velocity (three-fourths of that in the positive ion). It is, however, open to question whether this is the correct arrangement of the two electrons in the neutral atom. Each of them might also revolve in a separate orbit about the nucleus, and it rather seems that such a configuration could be more stable than that postulated by Bohr.

A helium nucleus with three electrons would require less energy to be expended in breaking it up than does one with two electrons and so cannot exist. There is no negatively charged helium ion. It also follows from this that a helium atom has no affinity for other elements, i.e. that helium is an inert gas. We shall return to this point when we discuss the formation of molecules.

3. LITHIUM

According to the theory there may exist, besides the nucleus with its three positive charges, an ion

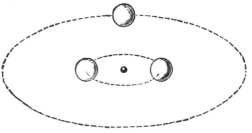

[FIG. 38.—Lithium atom.

with two or with one positive charge. The former ion contains an electron at a distance one-third of that in the hydrogen atom and having a velocity of revolution nine times as high ; the latter ion contains two electrons at a rather greater distance and with a rather smaller velocity of rotation.

A neutral lithium atom is probably constructed as shown in Fig. 38. Two electrons rotate on an inner orbit, a third electron, more weakly held, on an outer orbit. This single electron determines the monovalency of lithium. The radii of the inner and outer orbits are 0·362 and 1·182 times that for hydrogen. The number of revolutions per second is for the two inner electrons 7·65 times, for the outer one only 0·71 times, that of the hydrogen electron.

Under certain circumstances yet another electron may enter the outer orbit (which then becomes rather smaller) ; in this case we obtain a negatively charged lithium ion.

4. BERYLLIUM

With beryllium, which has four electrons, a neutral atom may be formed either by all four electrons moving in an orbit of radius 0·329 (taking that of hydrogen as unity) at equal distances from one another, or by two electrons revolving in an inner orbit of radius 0·262 and the other two in an outer orbit of radius 0·673 ; the former two would have an angular velocity 14·6 times, the latter two 2·2 times as great as that of the electron of hydrogen. Of these possibilities the second is the more probable, for beryllium is chemically divalent and the two outer, weakly bound electrons could produce this divalency.

Atoms with a Greater Number of Electrons.—The possible arrangements of electrons round a nucleus with a large number of positive charges are so numerous that there is at present no certainty or even any sharp criterion for one assumption or another. It is to be supposed that the electrons are arranged in rings or shells about the nucleus, but there is no certainty as to the number and composition of the rings or even as to whether the different rings lie in one and the same plane. We are mostly restricted to plausible assumptions, but these may lead to error just as well as to the truth. One clue is afforded by the periodic system of the elements, which shows that atoms in every ninth position are similar to those in the first position, i.e. that there is a periodicity with the number 8. To be more exact, our table on page 112 shows that there are the following differences in atomic number between the successive elements which come under one another in each column and are thus similar :—

$$8, \quad 8, \quad 18, \quad 18, \quad 32.$$

On the other hand, the difference between successive elements under a and b is—

$$10.$$

A second clue is given by the valency which, in such a series of 8 elements, increases from 0 to 4, and then as we go up the series falls again to 0. We are fairly well justified in ascribing the valency of the elements to 1, 2, 3 or more electrons lying on the outermost ring of the atom, these being the electrons by means of which the atom unites with another. On this view the rare gases would have no such free electron outside, since they have zero valency, but would end with a ring fully occupied. The further assumption is also plausible, that once such a complete ring has been formed in an atom it continues in the following atoms, so that the other electrons added form a new ring. It is, nevertheless, also possible that the stability of an internal ring may be altered by the addition of fresh electrons on the outside, and that another intermediate ring may be formed between them. It is probable, too, that the electrons are not arranged in concentric rings but in spherical shells.

But all these ideas are so far essentially nothing but suggestions for which there is no direct evidence. Various attempts have been made, first by Bohr himself, and then by Kossel, Vegard, Ladenburg, Dangmuir, and others, to represent the whole system of elements in this manner, by arrangements of rings with varying contents.

If the above-mentioned result of Debye (p. 147) and its extension by Kroo were to be accepted, namely, that the innermost ring in all atoms from sodium onwards consists of 3 electrons and the second always of 8 electrons, there would be definite evidence, at any rate for the first 20 to 25 elements. But the results of Debye and Kroo may be valid or they may not, and similarly in all these attempts to establish the arrangement of the electrons it is always a matter of guesswork. More accurate investigations on line-spectra and X-ray spectra may perhaps gradually lead to the possibility of increased certainty. In the meantime, however, it is needless to cite here the various mutually contradictory suggestions.

Just recently Bohr himself has gone into this question from the theoretical standpoint ; he insists that the configurations of the outer electrons are not produced all at once, but successively, according to the nearness of the electrons to the nucleus. On this view there can be only 2 electrons revolving on the one-quantum orbit, but there may be 8 on a two-quantum, 18 on a three-quantum, and 32 on a four-quantum orbit. The electrons belonging to each such orbit are dependent on, or *coupled* with, one another, and it is this coupling which determines the arrangement. The electrons which are coupled together may be moving on different orbits, and it is not necessary that the two-quantum orbits should be wholly inside the three-quantum one, but a two-quantum orbit may run outside an orbit of three or more quanta. According to this theory, the number of electrons in group O of the periodic system (the rare gases), starting from the inside, is as follows, the subscript figure indicating the quantum number of the orbit :—

Helium 2_1 (i.e. 2 electrons on a one-quantum orbit).

Neon 2_1, 8_2.

Argon 2_1, 8_2, 8_2.

Krypton 2_1, 8_2, 18_3, 8_2.

Xenon 2_1, 8_2, 18_3, 18_3, 8_2.

Radium emanation 2_1, 8_2, 18_3, 32_4, 18_3, 8_2.

But this theory again does not seem to lead to reliable results.

Possibly an investigation of the formation of molecules with this model of the atom may shed further light, but such an investigation appears to be very difficult and only the first beginnings of it have as yet been worked out.

In *molecules* we have no longer one central nucleus to deal with but there are several, at least two, around which the electrons move. A neutral molecule of hydrogen will consist of two nuclei, at a certain distance apart, and two electrons. Let us suppose two atoms like that of Fig. 34 to be at first a long way apart, and let us then by means of an external force cause one nucleus to approach the other. The electron which revolves about the one nucleus will be attracted by the other nucleus, and *vice versá*, and this attraction will increase as the nuclei come closer together.   On account of this attraction the orbits of the electrons will approach one another more rapidly than we make the nuclei approach one another by means of our external force.   When the two nuclei are at a certain distance apart, the orbits of the two electrons will coincide and form a single orbit in which the two electrons rotate about the axis joining the nuclei.   At this moment the hydrogen molecule has been formed. The repulsion between the two nuclei is then neutra-

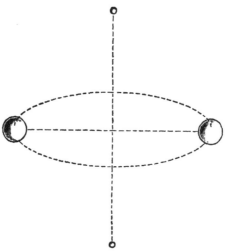

FIG. 39.—Hydrogen molecule.

lized by the attraction which the two electrons exert on each of the nuclei.

The combination of two atoms into a molecule arises, therefore, from the two electrons revolving in a common orbit about the line joining the nuclei. Accurate calculation shows that the radius of this orbit is rather smaller than that of the first orbit of a hydrogen atom (being 0·95 of it), and that the distance between the two nuclei is 1·16 times as great as the diameter of the latter orbit.   Fig. 39

gives an illustration of the hydrogen molecule. In absolute figures, the radius of the orbit is 0.52×10^{-8} cms. and the distance between the two nuclei 1.22×10^{-8} cms. Accurate numerical calculation shows, moreover, that the sum of the work requisite for the dissolution of the two atoms separately is less than that necessary for the dissolution of the molecule so formed. Thus the molecule is formed spontaneously from the atoms with evolution of energy which appears as the heat of formation of the hydrogen molecule.

It has always been a puzzle in the theory of atoms why two atoms of hydrogen should always unite closely to form a molecule, and what could be the property of the atom which occasioned this combination. We have here a simple and plausible explanation of it. The *common electronic orbit* is what produces the combination of two nuclei which would otherwise mutually repel one another.

We must not, however, omit to mention that objections may be urged on the score of stability against this conception of the molecule of hydrogen, objections which may perhaps require a modification of it. But the combination of the two nuclei, in such wise that the electrons rotate about the line joining them, will remain unaffected. It provides the only explanation of the formation of molecules.

According to this, one would at first expect that with the next most simple element, helium a similar union of two atoms to form a molecule could occur ; this is not the case, since helium is monatomic, as is well known. Suppose we proceed here in the same manner as we did with hydrogen and cause two helium atoms to approach one another in the direction of the line joining the nuclei. Each atom contains two electrons in one orbit about the nucleus (though we know from page 160 that this is not the only possibility for the helium atom). What will happen will be as follows :—

On account of the attraction of each nucleus on the electrons of the other atom the planes of the two electron orbits will approach one another more rapidly than the nuclei are brought together. At a certain instant it will happen that the two electron planes will coincide, while the nuclei are still at a certain finite distance apart. The four electrons in the same orbit will then, on account of their mutual repulsion, arrange themselves of their own accord so that each of them is a quarter of the circumference from the next. This appears to be a formation of a helium molecule. But if we now calculate the numerical value of the energy necessary for the dissolution of the two separate helium atoms and also the energy for the dissolution of the helium " molecule " thus resulting, we find that the latter is less than the sum of the former. Therefore, the *helium " molecule " cannot be formed* of its own accord. There must be an addition of energy from without in order to make helium atoms unite into a molecule. We may express this difference between the atoms of hydrogen and of helium by saying that two hydrogen atoms attract one another, whereas two helium atoms repel one another. This is the reason why helium atoms are monatomic. It is a problem for the future to examine similarly the formation of molecules from the higher atoms and it will be necessary to demonstrate that with neon, argon, xenon and krypton, in contrast to the elements near them, there is again this monatomicity.

Although the constitution of the higher atoms cannot be settled in individual cases, it may be postulated as probable that the production of a molecule from two atoms of the same or of different kinds takes place, in general, as it does with hydrogen, by the formation of a common ring about the two separate atoms ; this ring is what causes the combination of atoms into molecules. And, of course, it is to be assumed that in all molecules it is the outermost electrons, electrons of the outside ring, which so link up to form the molecule.

If we now consider the different effects of physical and chemical forces upon atoms according to these methods of representation, and examine how these effects are expressed, we have the following list :—

(1) Chemical phenomena take place essentially in the outermost rings of the atoms. These outermost rings fuse together and cause the production of the chemical molecule.

(2) The action of very high temperatures on the one hand, and of electrical excitement (in Geissler tubes) on the other, also affects the outer rings, and in many cases the inner as well, by breaking them up so that their re-formation produces the ordinary spectra.

(3) The bombardment of the electrons occasioned by X-rays influences principally the innermost rings of the atoms, which it destroys. By their re-formation the K-, L- and M-rays of the X-ray spectra result.

(4) Finally, the nuclei of the atoms are themselves concerned in radio-activity. The nuclei of the heaviest atoms disintegrate spontaneously and emit α- and β-rays. The concomitant γ-rays may, according to Rutherford, be regarded as the characteristic X-rays of the respective radio-active materials. The α-particles also decompose other nuclei, as has been shown with nitrogen.

Thus radio-acitivity invades the innermost part of the atom, its nucleus, and effects a real transmutation of atoms. For on the Rutherford-Bohr theory, a particular atom is characterized only by the charge of its nucleus. The number of electrons revolving about the nucleus depends on circumstances, and merely distinguishes between the neutral atom and the positive or negative ion of the atom. But a change in the nucleus produces a new atom. Up till now we have not been able to influence this change of the atom by any means at our disposal. Either it takes place of itself, spontaneously, as in radio-active substances, or it does not take place at all. By the new researches of Rutherford, which we have mentioned, the first step has been taken in the direction of interfering with it as we please. And if, as is to be hoped, this beginning is further developed, if we can use means to interfere with and to modify the changes, if we can make this disintegration of the nucleus quicker or slower and if we can extend it further to other atoms than hitherto, then this super-chemistry, the chemistry and physics of the nucleus of the atom, the manifold application of which to science we have here discussed, will have practical results beyond expectation.

INDEX